Springer Theses

Recognizing Outstanding Ph.D. Research

Aims and Scope

The series "Springer Theses" brings together a selection of the very best Ph.D. theses from around the world and across the physical sciences. Nominated and endorsed by two recognized specialists, each published volume has been selected for its scientific excellence and the high impact of its contents for the pertinent field of research. For greater accessibility to non-specialists, the published versions include an extended introduction, as well as a foreword by the student's supervisor explaining the special relevance of the work for the field. As a whole, the series will provide a valuable resource both for newcomers to the research fields described, and for other scientists seeking detailed background information on special questions. Finally, it provides an accredited documentation of the valuable contributions made by today's younger generation of scientists.

Theses are accepted into the series by invited nomination only and must fulfill all of the following criteria

- They must be written in good English.
- The topic should fall within the confines of Chemistry, Physics, Earth Sciences, Engineering and related interdisciplinary fields such as Materials, Nanoscience, Chemical Engineering, Complex Systems and Biophysics.
- The work reported in the thesis must represent a significant scientific advance.
- If the thesis includes previously published material, permission to reproduce this must be gained from the respective copyright holder.
- They must have been examined and passed during the 12 months prior to nomination.
- Each thesis should include a foreword by the supervisor outlining the significance of its content.
- The theses should have a clearly defined structure including an introduction accessible to scientists not expert in that particular field.

More information about this series at http://www.springer.com/series/8790

Maja D. Bachmann

Manipulating Anisotropic Transport and Superconductivity by Focused Ion Beam Microstructuring

Doctoral Thesis accepted by
The University of St Andrews, UK

Springer

Author
Dr. Maja D. Bachmann
Microstructured Quantum Matter
Max Planck Institute for Chemical
Physics of Solids
Dresden, Germany

Supervisors
Prof. Dr. Andrew P. Mackenzie
School of Physics
and Astronomy
University of St Andrews
St Andrews, UK

Prof. Dr. Philip J. W. Moll
Microstructured Quantum Matter
Max Planck Institute for Chemical
Physics of Solids
Dresden, Germany

ISSN 2190-5053 ISSN 2190-5061 (electronic)
Springer Theses
ISBN 978-3-030-51361-0 ISBN 978-3-030-51362-7 (eBook)
https://doi.org/10.1007/978-3-030-51362-7

This Springer imprint is published by the registered company Springer Nature Switzerland AG
The registered company address is: Gewerbestrasse 11, 6330 Cham, Switzerland

Supervisors' Foreword

One of the joys of supervising Ph.D. projects is seeing the number of different routes that can be taken to do outstanding work. Maja Bachmann is a true experimentalist in outlook. When approaching a new problem, she tackles it from every experimental angle that she can think of, performing exploratory research of a kind that too few people are prepared to attempt. Such a philosophy is only possible if one combines skill, intelligence and enthusiasm with a great deal of hard work. Maja is ideally suited to this approach. She is one of the most enthusiastic experimental physicists we have ever met, and just loves being in the laboratory, trying new things. As Maja demonstrated throughout her Ph.D., if someone as gifted as she is combines this approach with developing and refining entirely new experimental techniques, a fruitful route to new discoveries is established.

In this thesis, Maja describes examples of those discoveries, and the steps she needed to take to make them. She describes in detail how she refined focused ion beam sculpting of exotic compounds into a kind of art form, producing experimental devices of beauty as well as utility. She then goes on to recount two discoveries, those of strongly directional ballistic transport in the ultra-pure delafossite metals and remarkable patterns of superconductivity following complex strain profiles in micro-devices of $CeIrIn_5$. In both cases, the phenomena that needed to be explained resulted from observations that could be described as serendipitous. They were serendipitous in the sense that they were unexpected, but they were not lucky. Maja made her own luck by developing such a high level of skill, and being prepared to try many investigations of the materials of interest. In a world where 'experiments' are too often more like attempts to verify pre-existing theories or prejudices than genuine explorations, it is refreshing to see this approach.

Once Maja had made interesting and unexpected observations, she entered 'detective mode' and hunted down the underlying physics. She combined a programme of follow-up experiments with theoretical simulation and excellent physical intuition. Some of the simulation work was done in collaboration with other groups, but most of the key intellectual breakthroughs came from Maja personally.

It takes an unusual combination of talents to work in this way, at such a high level, and then write it up in an outstanding thesis like this.

Maja was both a pleasure to work with and a colleague with the precious ability to inspire. It was a privilege to supervise her, and we look forward to observing her go from strength to strength as her research career progresses.

Dresden, Germany
March 2020

Prof. Andrew P. Mackenzie
Prof. Philip J. W. Moll
(joint supervisors)

Abstract

This thesis presents the results of electrical transport experiments performed on two microstructured quantum materials, namely, on the ultra-pure metal $PdCoO_2$ and on the heavy fermion superconductor $CeIrIn_5$. Throughout this work, focused ion beam (FIB) microsculpting was utilised to design the investigated devices.

I begin with an introduction to the FIB instrument, with a specific focus on its application for microstructuring transport devices from quantum materials. In particular, our standard fabrication procedure, in which a thin slab of material is extracted from a bulk single crystal for further processing is described in detail, as this approach can be utilised for most metallic compounds. Furthermore, I describe a micro-fabrication process for creating transport devices from platelet-shaped single crystals.

Thereafter I present ballistic transport measurements of the ultra-pure delafossite metal $PdCoO_2$. By investigating mesoscopic transport bars which are narrower than the electron mean free path (up to 20 μm), I demonstrate that the ballistic transport in $PdCoO_2$ is strongly anisotropic as a result of the underlying quasi-hexagonal Fermi surface shape. Moreover, I report on the results of transverse electron focusing (TEF) experiments, a technique which directly probes the real space ballistic trajectories of electrons in a magnetic field, which demonstrate the super geometric focusing effect.

Furthermore, by investigating microstructures of the superconducting heavy fermion compound $CeIrIn_5$ by means of transport measurements as well as scanning SQUID microscopy in collaboration with external groups, a route to controllably manipulate the local strain in microstructured devices was found. The presented approach is based on exploiting the substrate-induced biaxial strain due to differential thermal contraction, which is spatially tailored by defined FIB cuts. As the superconducting transition in the heavy fermion compound $CeIrIn_5$ is highly sensitive to strain, the local T_c within the device is controlled via the spatial strain distribution.

Publications Related to This Thesis

[1] M. D. Bachmann, N. Nair, F. Flicker, R. Ilan, T. Meng, N. J. Ghimire, E. D. Bauer, F. Ronning, J. G. Analytis and P. J. W. Moll. Inducing superconductivity in Weyl semimetal microstructures by selective ion sputtering. *Science Advances*, **3**:5, 2017.

[2] M. D. Bachmann, G. M. Ferguson, F. Theuss, T. Meng, C. Putzke, T. Helm, K. R. Shirer, Y. S. Li, K. A. Modic, M. Nicklas, M. König, D. Low, S. Ghosh, A. P. Mackenzie, F. Arnold, E. Hassinger, R. D. McDonald, L. E. Winter, E. D. Bauer, F. Ronning, B. J. Ramshaw, K. C. Nowack and P. J. W. Moll. Spatial control of heavy fermion superconductivity in $CeIrIn_5$. *Science*, **366**:6462, 2019.

[3] M. D. Bachmann, A. L. Sharpe, A. W. Barnard, C. Putzke, M. König, S. Khim, D. Goldhaber-Gordon, A. P. Mackenzie and P. J. W. Moll. Super-geometric electron focusing on the hexagonal Fermi surface of $PdCoO_2$. *Nature Communications*, **10**:5081, 2019.

Acknowledgements

I would like to thank Philip Moll for guiding me through this Ph.D. and teaching me the art of FIB microstructuring. I truly enjoyed Philips' drive, curiosity and never ending creativity, which have greatly influenced this work. Moreover, I appreciated the many frivolous and serious discussions we shared over plates of outstanding meals in fantastic locations around the globe.

I am also grateful to Andy Mackenzie for giving me the opportunity to pursue my Ph.D. at both the University of St Andrews in Scotland and the Max Planck Institute for Chemical Physics of Solids (MPI-CPfS) in Dresden. I always felt welcome to walk through his open office door and discuss all aspects of both physics and life regardless. The friendly and fun, yet serious, atmosphere he created has made working at the institute enjoyable every day.

The work presented in this thesis would not have been possible without the help and support of countless collaborators. Nityan Nair, James Analytis, Roni Ilan, Tobias Meng, Nirmal Ghimire, Eric Bauer and Filip Ronning have been essential for realising my work on NbAs. In particular, Eric and Filip have grown numerous high-quality crystals, including $CeIrIn_5$, which were the starting point of many of my explorations of microstructures. Further, I would like to express my gratitude towards Ross McDonald and Fedor Balakirev from the Los Alamos NHFML, who supported me during my pulsed field measurement experiments. Furthermore, the in-house collaboration with Elena Hassinger, Frank Arnold and You-Sheng Li, as well as with Matt Ferguson, Florian Theuss, Katja Nowack and Brad Ramshaw from Cornell University was vital in understanding our initially puzzling results in $CeIrIn_5$, presented in this thesis. Moreover, I would like to thank Pallavi Kushwaha and Seunghyun Khim for growing the world's cleanest oxide metal, $PdCoO_2$, which enabled the ballistic transport studies presented here. The discussions with Aaron Sharpe, Arthur Barnard and David Goldhaber-Gordon as well as with Thomas Scaffidi in conjunction with their simulations greatly aided the understanding of mesoscopic transport in $PdCoO_2$. Lastly I would like to thank Arkady Shekhter for sharing his thoughts and ideas during our discussions of a wide range of condensed matter systems throughout the past years.

As a student of the Scottish Centre for Doctoral Training in Condensed Matter Physics (CM-CDT), I had the opportunity to spend nearly a year in St Andrews for which I am very grateful. A broad range of taught courses and transferable skill events broadened my scientific understanding and expertise. At this point, I would like to thank the administrative staff, Christine Edwards, Julie Massey, Wendy Clark and Debra Thompson for their professional support and assistance.

That a town is only as good as the people in it has become clear to me during my time in St Andrews thanks to Scott 'the crab' Taylor, Matt Neat, Elliott Levi, Lewis Bawden, Oliver Clark, Rhea Stewart, Ian van Beek, Artem Strashko and Mike Lynch-White. Thank you for sharing the British way of Ph.D. life with me. I will always look back with a smile on a my face thinking about the bonfire nights on East Sands, the fierce competitions and engaging discussions in room 120, our $\frac{2}{3}$ pints at StAbco, the afternoon strolls to Jannetta's and many more nice memories. Special thanks to JPh Reid for teaching me how to conduct thermal transport measurements and introducing me to the joys of proper alpine cycling. I wish I could have spent more time exploring the hills of Scotland with Igor Marković and Steffi Matern, who moved into good old Chamberlain street just it was time for me to leave the Auld Grey Toon.

I am very fortunate to have been able to conduct my experimental research at the Max Planck Institute for Chemical Physics of Solids in Dresden, which has spoiled me with its fantastic research infrastructure. In particular, I would like to thank Markus König for his effort in maintaining the FIB instruments and Sebastian Seifert for his support in the clean room. Further I would like to express my gratitude to Dmitry Sokolov for teaching me how to use the Laue instrument. Moreover, I am grateful for the fantastic work and support I received from the technical staff from the mechanical and electrical workshop, as well as from the cryogenics team. Finally, I would like to thank Burkhard Schmidt for his IT support.

Next, I would like to thank the members of the microstructuring quantum matter group, MQM: Toni Helm, Kim Modic, Tino Zimmerling, Kent Shirer, Amelia Estry, Jonás Díaz, Xiangwei Huang and Carsten Putzke. I enjoyed sharing the lab benches with them just as much as sharing the dinner table at Salavatore's or the BBQ donut ride down the Elbe. I would especially like to thank Carsten for his help during the final year of my Ph.D. It has been an absolute pleasure to work with him and learn from his professional, skilful and thorough approach to experimental physics.

Life in Dresden would not have been nearly as enjoyable without Mark Barber, Jack Bartlett, Dan Brodsky, Cliff Hicks, Fabian Jerzembeck, Markus König, You-Sheng Li, Philippa McGuinness, Nabhanila Nandi, Hilary Noad, Alexander Steppke, Veronika Sunko, Eteri Svanidze, Elina Zhakina and Lishan Zhao. I am grateful for the fond memories of the fun times at the institute, the Friday evenings in Neustadt or in the occasional hiking outings to Sächsiche Schweiz. I will forever treasure the moment Nabhanila and I first realised that, indeed, *life is good*. We further continued to appreciate all aspects of life during our road trips with

Veronika on Vancouver Island, and in Scotland, Croatia and Switzerland as well as our occasional visits to SonderBar.

Lastly, my Ph.D. would definitely not have been nearly as much fun without Veronika! There is nobody I would have rather shared a flat, lunch, a post QFT waffle, and my thoughts and complaints with, than her. I am grateful for having met an extraordinarily kind, talented and FUN friend with whom I had the pleasure to share much of my Ph.D. experience with.

Finally, I am eternally grateful to my parents, Linde and Dino, for their unconditional love and support to help me reach where I am today. Moreover, I would like to thank my brother Mario and my grandma Emi for their encouragement to achieve my goals.

I gratefully acknowledge financial support from the Engineering and Physical Sciences Research Council via the Scottish Condensed Matter Centre for Doctoral Training under grant no. EP/L015110/1, and from the Max Planck Society.

Research data underpinning this thesis are available at https://doi.org/10.17630/38c95513-b893-4cc6-8f91-529100888e58.

Contents

Chapter 1
Introduction

Fundamentally, all materials are quantum mechanical by their very nature. This insight stems from the fact that the physical description of how atoms bond and electrons interact is provided by quantum mechanics. These effects are largely understood and in many cases can be approximated by a semi-classical description on a macroscopic level. In contrast, the term *quantum matter* refers to a vast class of compounds where quantum effects play an important role leading to novel, and often unexpected, electronic behaviour. These tend to be materials in which the electrons can no longer be considered independent, but interact strongly giving rise to collective excitations and emergent phenomena, and hence are commonly termed *strongly interacting electron systems*. Examples of quantum materials, several of which will be investigated in this thesis are superconductors, graphene, topological insulators, Weyl semi-metals, heavy fermion metals, quantum spin liquids and spin ice systems.

A primary aim of condensed matter physics is to understand and ultimately control the rich variety of collective behaviour resulting from the interactions amongst electrons in a solid. This quantum many-body problem poses both an intriguing intellectual challenge, while also forming the basis for quantum applications, for instance quantum computing or quantum sensing functionalities. To these ends, incredible effort has gone into synthesising compounds and subsequently characterising and comprehending their properties—and occasionally even detecting new phases of matter. For such endeavours there are a plethora of tools and techniques a condensed matter physicist may employ to investigate materials of current interest. Some of which probe the thermodynamic state or free energy of a system directly while others examine related properties, for instance electronic transport, which is the focus of this work.

Arguably one of the biggest challenges in condensed matter physics concerns the purity and homogeneity of these often complex materials. In particular, the quantum states are typically delicate and sensitive to impurities and disorder. Furthermore, when materials are first synthesised, they are often only available as microscopic and

M. D. Bachmann, *Manipulating Anisotropic Transport and Superconductivity by Focused Ion Beam Microstructuring*, Springer Theses,
https://doi.org/10.1007/978-3-030-51362-7_1

Fig. 1.1 Scanning electron micrograph of a $AgNiO_2$ crystallite which was microstructured into a transport bar, in which the longitudinal and transverse voltage can be measured simultaneously. Image kindly provided by C. Putzke

irregular crystallites or polycrystals. In these cases performing a traditional electronic transport measurement is challenging as there is only little control over the shape and alignment of the crystal. Further, ensuring a homogenous current flow along a specific crystalline axis is a demanding task, which is why the crystals are frequently wire sawed and/or polished into rectangular bars. Mechanically ablative procedures, however, are not suitable for sample sizes below $\sim 30\,\mu m$ and additionally may introduce undesired strain and dislocations into the crystal. Lastly, even once all these obstacles are overcome successfully, often one is left with small signal sizes due to a sub-optimal sample aspect ratio.

Recently, however, focused ion beam (FIB) microstructuring has created many new and exciting opportunities for investigating and exploiting quantum materials. This comparatively young technique has just started to revolutionise traditional sample fabrication methods by offering the ability to electrically contact and micro-sculpture microscopic samples, and there remains much to explore. In essence, the FIB is a tool with which matter can be precisely removed locally, while preserving the crystalline quality of the remaining material. In particular for fabricating electrical transport devices, this addresses many challenges at once. Figure 1.1, for instance, displays a tiny crystal platelet which is too small to contact by hand using conventional four-point wiring. Utilising the FIB, however, it is straightforward to create a device in which two longitudinal and two transverse voltages can be measured along a well-defined current path. Similarly, large but inhomogeneous compounds can be characterised by probing only small, selected regions within the sample. Moreover, the ability to microscopically manipulate the sample also provides control over the crystal orientation, such that the resistivity can be accurately determined along any direction.

Microstructured samples, however, not only provide access to resistivity measurements in small or peculiar shaped crystals. They also offer the possibility of investigating mesoscopic and microscopic transport regimes, which are otherwise difficult if not impossible to achieve in many exotic materials. This concept is visualised in Fig. 1.2. In general, L represents a characteristic length scale of the system, for example the coherence length of a superconductor or charge/spin density wave, the typical distance between domain walls (structural, magnetic), or the electron

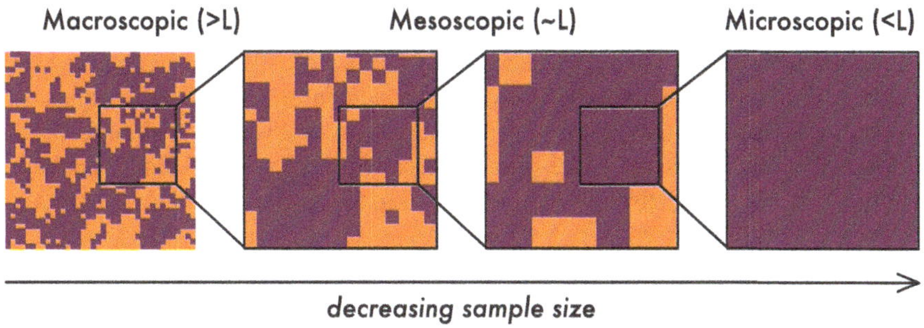

Fig. 1.2 By tuning the sample dimensions across an intrinsic, characteristic length scale L of the system, various physical regimes can be realised. Between these regimes, the qualitative behaviour of the system can be drastically different. This figure is based on an illustration by P. Moll

mean free path. At the same time, the device sizes achievable with the FIB range from millimetres down to several hundred nanometers, spanning over four orders of magnitude in size. Thus by tuning the sample dimensions appropriately, a qualitative difference in the behaviour of the system may be expected when crossing from the macroscopic to the microscopic regime.

Additionally, once a system has entered the mesoscopic or microscopic regime, the sample shape and dimensions become relevant parameters for the observed behaviour. In turn, the form of a device can be employed to spatially manipulate and influence its electrical characteristics. Throughout this thesis *functionality by form* will be a recurring theme, when studying ballistic effects in $PdCoO_2$ in various device geometries or during the investigation of the spatially modulated superconductivity observed in $CeIrIn_5$.

However, as FIB microstructuring of quantum matter has only been developed within the last decade, it is still a relatively new technique with much unexplored territory and a host of open questions. In this respect, I think of the work presented in this thesis as an adventurous approach towards microstructuring to find out what is possible. During this exploration, I investigated the effects of FIB irradiation on an ultra-pure metal as well as on a family of mono-arsenide compounds, in which surface superconductivity can be induced. In addition, I studied the role of substrate strain on microstructured devices as a result of puzzling resistivity anisotropy measurements. These experiments showcase the capabilities of the FIB to manipulate quantum materials and create devices in order perform measurements which are otherwise not possible.

The three main chapters that make up this thesis are designed to be self-standing. Each contains an individual introduction to the relevant background physics and a literature reviews as well as research results. They can therefore be read independently of each other, and are structured as follows:

- In Chap. 2 I provide an overview of the FIB machine and its operating principle. In particular, I have focused on describing the use of FIBs for microstructuring quantum matter transport devices. My target audience is future FIB enthusiasts, who will hopefully find some helpful descriptions and recommendations on the following few pages. A reader purely interested in the experimental results on either $PdCoO_2$ or $CeIrIn_5$ can comfortably skip this section, and move forward to the respective chapter.
- In Chap. 3 I describe work investigating in-plane ballistic transport in the layered delafossite metal $PdCoO_2$. The mean free path in this compound is up to 20 μm below 20 K—a length scale which can be reached and manipulated with the FIB. As a result, ballistic devices could be fabricated and revealed novel insights into the direction-dependent ballistic in-plane transport in $PdCoO_2$.
- Chapter 4 is devoted to the heavy fermion compound $CeIrIn_5$. Because the superconductivity in $CeIrIn_5$ is particularly sensitive to uniaxial pressure, it was found that the substrate-induced strain in the microstructure due to differential thermal mismatch is enough to substantially modulate T_c in a spatially anisotropic manner. This result provides a new route to engineer potential landscapes for electrons in strongly correlated matter, and finally solves the longstanding mystery of the anomalously high resistive transition in $CeIrIn_5$.

Chapter 2
Focused Ion Beam Micro-machining

Focused ion beam (FIB) micro-machining has become an invaluable tool for studying quantum materials due to its capability to precisely define the shape and dimensions of single crystals on the sub-micrometer scale. In particular, high-quality mesoscale electrical transport devices can be fabricated from newly synthesised compounds or materials which are too small to manipulate otherwise. Moreover, by designing the microstructure accordingly, one can enhance the experimentally accessible signatures of the sought after physical phenomena. Further, this then enables prototype devices to be directly produced on standardised chips in order to investigate the application potential of a material for future electronics.

In the following section, I will first review the standard operating principle of FIB/SEM dual-beam systems, followed by a comparison of the predominantly used gallium-based ion beams with the more recently (2006) developed xenon plasma-based FIBs. However, this chapter is not focused around the technical details of the FIB, but is instead intended to introduce the relevant ion beam fundamentals in order to apply the FIB for microstructuring of quantum matter. Hence the majority of the following chapter will cover essential information for fabricating functioning transport devices. In particular, the second half of the chapter is devoted to giving a step-by-step guide on how to fabricate electrical transport devices from lamellae and platelet shaped single crystals.

2.1 Fundamentals of Dual-Beam SEM/FIB Systems

At the time when the first scanning *electron* microscope (SEM) systems became commercially available in 1965, the idea emerged to also utilise *ions*, which were heavier and therefore more energetic than electrons, to image and manipulate matter on microscopic length scales. The early ion beam sources were based on field emission technology, which were low in brightness and suffered from position sta-

M. D. Bachmann, *Manipulating Anisotropic Transport and Superconductivity by Focused Ion Beam Microstructuring*, Springer Theses,
https://doi.org/10.1007/978-3-030-51362-7_2

bility issues. The breakthrough came in 1975 with the invention of the liquid metal ion source (LMIS), which is stable over time and offers a high brightness. Their subsequent incorporation with scanning electron microscopes in 1979 formed the foundation of modern dual-beam FIB/SEM systems.

Today, the FIB has become a commercially available tool, which is invaluable for microanalysis and microstructuring applications. It is heavily used in the semiconductor industry for circuit editing, failure analysis and quality control of micro-chips on the single transistor level. In materials science, the FIB is commonly used for the preparation of thin slices of a material, so-called lamellae, for investigations with transmission electron microscopes (TEM). The FIB's cross sectioning capability is further exploited in biological applications, in which biological tissue can be sectioned and imaged (so-called *slice-and-view*) in order to create a reconstructed 3D model. Moreover, nano-fabrication of MEMS (micro-electro-mechanical systems) devices, in which both the sputter and deposition abilities of the FIB are exploited, are a further application of the FIB.

For the purpose of microstructuring quantum materials, the FIB is an ideal tool to sculpt materials with a high level of precision to create three dimensional structures from almost any metallic compound. With a dual-beam system, the samples can be imaged non-destructively with the SEM and locally manipulated (material removal or deposition) by means of the energetic ion beam. A generic and simplified dual-beam system is sketched in Fig. 2.1a. The gallium-based *Helios NanoLab G3 CX* FIB, which was the workhorse of this thesis, is displayed in panel b. The second, xenon-plasma-based FIB system (*Helios G4 PFIB UXe*) used for this work is nearly identical in size.

The sample is mounted on a 5-axis stage, which allows it to be moved in the (x, y)-plane of the sample chamber, rotated around the vertical axis of the stage and tilted towards the ion beam column. With this, the region of interest on the

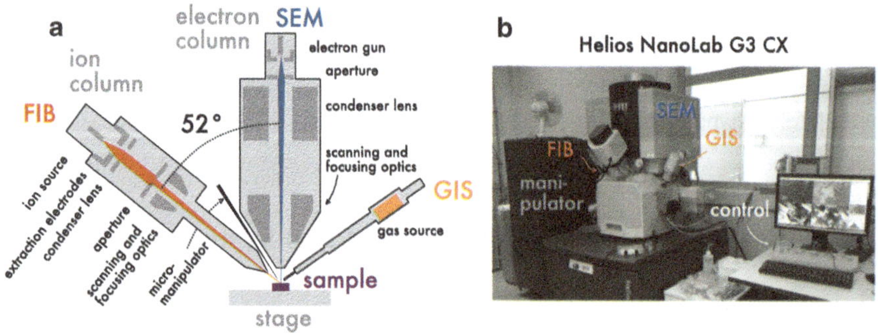

Fig. 2.1 **a** Schematic illustration of a simplified dual beam SEM/FIB machine. The angle between the electron and ion column is typically around 50°. The sketch is based on [1]. **b** Optical image of the gallium based *Helios NanoLab G3 CX* system at the Max Planck Institute for Chemical Physics of Solids in Dresden

sample is then moved into the coincidence point of both beams, such that it can be imaged with both the FIB and SEM, yielding stereo vision for the user.

The ion beam column is typically tilted around 50° degrees with respect to the vertical electron column. Its main components are an ion source, an acceleration stage and electrostatic lenses to scan and focus the beam. By far the most common ion source technology is based on liquid gallium due to its low melting point. Recently, xenon plasma FIBs have also become commercially available. The differences between these source technologies is discussed in the next section. After exiting the source, the ions are accelerated with a voltage typically in the range between 1 and 30 kV, yielding ions with energies up to 60 keV (for Ga^{2+} and Xe^{2+}). Subsequently a set of apertures limits the beam diameter and hence current of the ion beam to the value selected by the user. Finally, the ion beam passes a set of electrostatic lenses allowing it to be raster scanned and focused onto the sample.

Similarly, the SEM raster scans a beam of electrons across the sample, providing non-invasive, high-resolution images of the sample. Moreover, the electron column can be operated as stand-alone instrument allowing EDS (energy-dispersive X-ray spectroscopy) and EBSD (electron backscatter diffraction) analyses to be performed.

The gas injection system (GIS), sketched in orange in Fig. 2.1, offers the capability to locally deposit metallic or insulating material on the sample of a surface by means of ion-assisted chemical vapour deposition. This process is described in detail in Sect. 2.1.4.

Lastly, many FIBs are equipped with an integrated micro-manipulator consisting of a sharp tungsten needle, which can be moved in three dimensions and rotated around its axis. In combination with the welding capabilities offered by the GIS, this allows small samples and lamellae to be manipulated and transferred within the FIB.

2.1.1 Ion Sources: Gallium Versus Xenon Plasma

Liquid metal ion sources (LMIS) are by far the most used ion beam sources due to due to their ability to self-form a sharp tip, yielding a position-stable source with high brightness. Typically gallium is utilised for this purpose because of its low melting point of around 30 °C and the fact that Ga easily forms a super-cooled liquid at room temperature. Moreover, a low vapour pressure renders gallium compatible with high-vacuum (HV) applications, and its low surface free energy promotes viscous behaviour on the needle substrate. Many other metals and alloys have successfully been exploited as well, but gallium has become the industry metal of choice. Ultimately the material removal rate of the liquid metal source FIB technology is limited by the eventual break-down of the self-formed tip at high extractor voltages. Hence, a different source technology has been developed with the aim of achieving enhanced milling rates. Xenon based FIBs operate by generating a xenon plasma, from which ions are extracted through an aperture. These fundamentally different ion emission strategies lead to distinct beam profiles, which one should be well aware of when microstructuring, as they influence the milling properties. To this end I will dis-

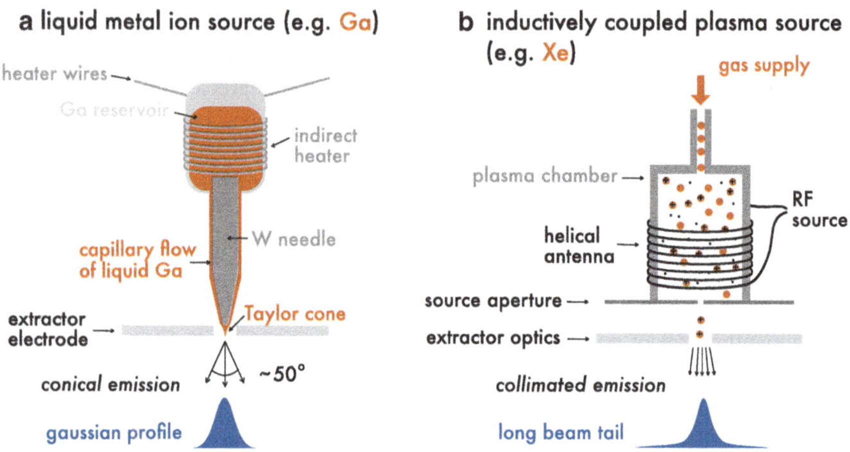

Fig. 2.2 **a** Sketch of a gallium-based liquid metal ion source (LMIS). A coil heater ensures that the gallium in a reservoir is molten, such that it can flow along and wet the surface of a connecting tungsten needle. At the end of the needle a cone-like tip (the *Taylor cone*) is self-formed as a result of the applied electric field from the extractor electrode. If a critical voltage is exceeded, ions are emitted from the tip and lead to a stable gaussian beam profile. **b** A plasma FIB is based on inductively coupling a radio frequency (RF) antenna to a reservoir of xenon gas. This forms a dense plasma, which can be extracted through a source aperture. This forms a collimated emission pattern with high angular intensity, which results in a significantly larger beam tail compared to the GFIB. Sketch based on [2]

cuss the fundamentals of LMIS and inductively coupled plasma (ICP) sources in the following section.

A sketch of a characteristic gallium-based liquid metal ion source is presented in Fig. 2.2a. A reservoir is heated to above the melting point of gallium, such that the liquid metal can flow along the tungsten needle which is attached to the reservoir and wet its surface. The end of the tungsten needle, which has a sharp tip (~2–5 μm radius), is then brought near the electric field created by an extractor electrode. The solution of the hydrostatic equations balancing the electrostatic forces against the surface tension of the liquid metal yields a self-formed tip ~2–5 nm diameter), the so-called Taylor cone. When the voltage on the extractor electrode exceeds a few kilovolts, Ga^{2+} ions are emitted in a conical fashion leading to a gaussian shaped beam profile. Crucially, for the spatial resolution of the FIB, the liquid gallium at the tip of the tungsten needle must retain a stable position in space, such that the electron-optical elements in the ion column can precisely steer and focus the ion beam. For this, the ability of gallium to remain a super-cooled liquid at room temperature is essential as it allows the heater to remain inactive during operation. In contrast, in non-gallium-based LMIS the convection in the continuously heated liquid metal sources greatly impairs the stability of the Taylor cone, resulting in an inferior spatial precision of the ion beam.

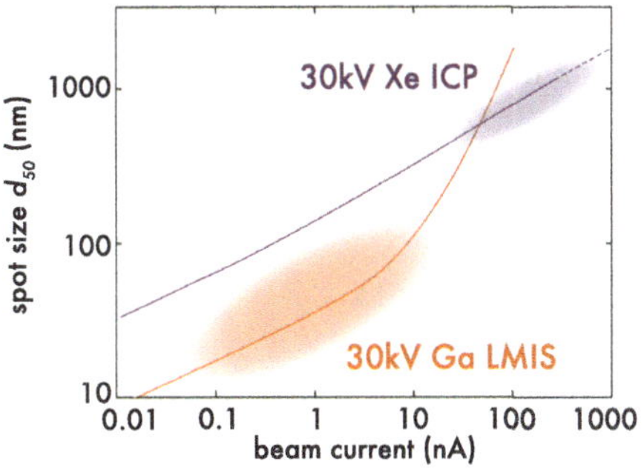

Fig. 2.3 Comparison of the spot sizes of gallium based liquid metal ion sources (LMIS) to xenon inductively coupled plasma sources (ICP) as a function of beam current [3]. The spot size d_{50} denotes the effective central beam diameter containing 50% of the ion current. Above ~10 nA spherical aberrations degrade the performance of the GFIB (loss of focus). Hence at high beam currents the PFIB yields smaller spot sizes, in particular in state-of-the-art plasma FIBs with emission currents above 1000 nA (dashed purple curve). Figure based on [3]

The typical spot size as a function of emission current of a gallium FIB is presented in Fig. 2.3. At low currents, Coulomb interactions within the beam limit the spot size to less than 100 nm. However, the relevant parameter for milling microscopic structures is not the minimum size of the beam diameter, but instead the sharpness of the edge of the ion beam. The more steeply the ion beam profile drops off at its periphery, the sharper the boundaries of the milled structures will be. At the other end of the spectrum, above ~10 nA spherical aberration effects additionally degrade the performance of the beam, as indicated by the change in slope of the spot size as a function of beam current. The maximal beam current achievable with the LMIS technology is ~100 nA, after which the formation of the Taylor cone becomes unstable.

In 2006, the design of xenon plasma based ion sources opened the route towards high current FIB milling enabling the use of up to several μA of ions, while maintaining a high beam quality and performance. This is possible due to the greater collimation of the ion beam produced by the plasma FIB, even at high currents. A sketch of such a system is displayed in Fig. 2.2b. The plasma chamber, filled with xenon gas, is coupled to a radio frequency antenna, which is helically wrapped around the chamber. The induced electric field sufficiently accelerates the electrons to ionise the atoms and thereby creating a dense xenon plasma. Finally, the ions are accelerated by an extraction electrode and collimated through an aperture. The profile of the ion beam created in this fashion maintains a power law emission, even at high currents, and thereby exceeds the performance of the GFIB for currents greater than ~100 nA, c.f. Fig. 2.3, purple curve.

Finally, another relevant feature, especially for materials science, is the non-metallic and inert character of xenon ions, which prevents chemical interaction with the target material, such as unwanted formation of metallic compounds on the surface of the microstructured metals. Moreover, as the implanted xenon ions do not react with the target material and have a large radius which prevents them from diffusing into the sample, the xenon in the surface layer of the irradiated material can escape relatively easily [4].

2.1.2 The FIB Milling Process

One of the main functions of a FIB is the local removal of material by means of ion sputtering. On a microscopic level this process is a consequence of the loss of the ion's kinetic energy through the interaction with the atoms of the target material upon impact with a solid. In general, a number of different processes occur when an ion interacts with a solid, including ion reflection and backscattering, electron emission, atomic sputtering and sample heating. In particular surface ablation takes place if the energy transferred to an atom near the sample surface exceeds the critical *displacement energy*, such that a recoil atom may be emitted from the solid [5]. Over time this generates a spot-sized hole, which can range from sub-100 nm to tens of micrometers in size. The ion beam can then be steered using electrostatic lenses, enabling it to be raster scanned across the surface of a sample, such that arbitrarily shaped holes of variable depth can be milled. On the level of a FIB user, there are number of factors, such as the beam diameter, scan direction and order, which can be controlled in order to optimise the milling process, both in terms of speed and cleanliness, which will be the subject of the following section, while keeping in mind the particular aim of fabricating transport devices.

There are a number of requirements which must be fulfilled in order to fabricate correctly functioning microstructured transport devices:

- The bulk of the sample must not be damaged.
- Short circuits must not exist. Erroneous electrical connections distort the current flow and lead to an incorrect resistance measurement.
- The geometry and dimensions of the sample must be well known in order to determine the geometric factor precisely, such that the resistivity can accurately be calculated from the measured resistance.
- Parallel conductances must be minimised. Using a FIB it is unavoidable to create a damage layer around the sample, which can be regarded as a parallel resistor. If its resistance is high enough however, it can be neglected.

In order to best meet these demands, it is essential to understand the FIB milling behaviour and its consequences. Figure 2.4 graphically summarises the fundamentals of FIB patterning based on the xT software of FEI used for this work (version 10.1.3). Other FIB software offer comparable tools. The essential aspects portrayed in the following section, however, are generally applicable and independent of the software in use (unless explicitly stated otherwise). As mentioned above, the ion beam is raster scanned, c.f. Fig. 2.4a. This implies that the beam is focused onto an individual spot for the duration of the so-called dwell time, $t_{\mathrm{dwell}} \sim \mu\mathrm{s}$, before it progresses to the next spot. The sputtering rate is typically non-linear as a function of dwell time. Initially,

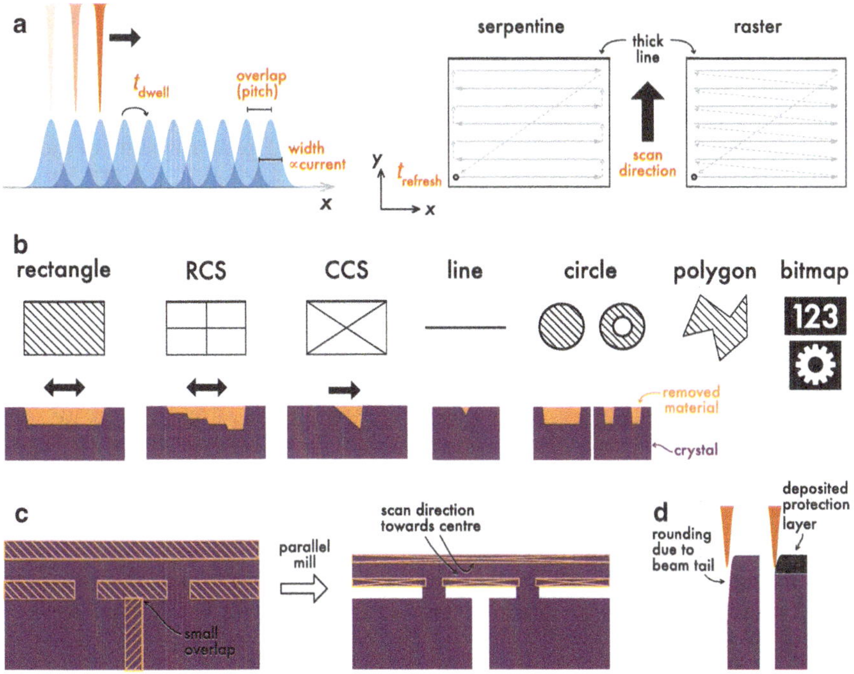

Fig. 2.4 The essentials of FIB patterning described on the basis of FEI's xT software package. The same or similar functionalities are available in other FIB software tools. **a** The most commonly used *scan types* for rectangular shaped patterns are *serpentine* and *raster*. In both cases, the scan generator sweeps the beam in the scan direction and has the option to pause for a time t_{refresh} after each cycle. Every line consists of a number of beam 'pixels', which are separated by a certain *pitch*. At each pixel the beam remains for a time t_{dwell} before moving to the adjacent pixel. **b** Top row: The basic patterning shapes available in FEI's xT user interface. See main text for detailed description. Bottom row: Cross section of the respective patterns. The orange areas represent the volume which will removed by the respective pattern. **c** Recommended milling strategy for microstructuring. Here a section of a four-point bar is displayed. The cleanest structures are typically achieved by exploiting the optimised scan strategy for rectangular patterns in combination with the parallel milling mode. If necessary, the side walls of the structure can be polished with a cleaning cross section. **d** Due to the beam profile, the top edge of a CCS polished side-wall is typically rounded. This effect can be avoided by depositing a protection layer

as the dwell time increases, the material locally heats up more, which facilitates atom ablation [6]. At the same time, redeposition effects limit the sputtering rate above a critical value [7]. The horizontal and vertical distances between the individually milled spots are separated by the so-called x and y pitch, which are typically defined as a function of beam diameter, which itself is a function of the beam current. For instance, a standard rectangular pattern is milled with *50% overlap*, indicating that the individual gaussian shaped beam profiles overlap by 50%. Commonly used scan strategies for milling are the so-called *serpentine* or *raster* patterns, displayed in panel a. They have a well-defined scan direction (towards the thick line, as indicated in panel a) and can be set to pause for a time $t_{refresh}$ after every complete pass (typically set to 0 for milling). Their essential difference is the scanning path, which can lead to a difference in surface morphology, sputter yield and artefact formation [8].

Similarly, the scanning path used to generate other milling patterns crucially influences the sputter yield, redeposition distribution and general quality of the cut. Panel b of Fig. 2.4 presents seven available patterning options in the FEI xT user interface, which I will describe in the following.

Rectangle pattern This is likely the most commonly used pattern and, in general, the preferred choice when microstructuring transport samples. Due to its scanning strategy (serpentine or raster) the *layer-by-layer* patterning yields a cross section with a flat bottom surface. Moreover, because the ion beam passes over the same area many times, the redeposition is comparatively low for shallow cuts (c.f. Fig. 2.5a). The standard setting is 50% beam overlap at a dwell time of 1 μs, which yield an evenly milled rectangular hole. However, if the pitch and dwell time are increased (e.g. −300%, 1 ms) this will instead create an array of spots, which were for instance used to create the 'roots' presented in Fig. 2.17.

Regular cleaning cross section (RCS) This pattern is designed for the fast removal of large volumes of material. The cross section resembles a stair case and is ideally suited for excavating lamellae. The standard setting is 50% beam overlap at a dwell time of 1 μs.

Cleaning cleaning cross section (CCS) Unlike the rectangle and the RCS pattern, where the beam passes over the same area multiple times, the CCS pattern operates on a *line-by-line* basis. Here, a single line is rastered multiple times, before the beam advances to the next line in the direction of the scan. However, once advanced, the beam never returns to a previously milled line (unless manually set to do so). The advantage of this strategy is to create side walls with minimal redeposition (c.f. Fig. 2.5b), hence this is often referred to as side wall *polishing*. For microstructuring, CCSs are essential to minimise redeposition and are used for polishing the side walls of the lamellae as well as the side walls of the finished structure. The standard beam overlap is 85% at a dwell time of 1 μs. Here, the increased beam overlap compared to the commonly used value of 50% creates an improved side wall finish due to the higher irradiation density which prevents uneven, serrated boundaries.

Line pattern The line pattern consist of a single line of beam spots. For microstructuring applications, the use of line patterns should generally be avoided. Instead rectangular patterns are recommended, as they create a cleaner cut, which greatly reduces the chance of accidental short circuits.

Circle pattern The circular pattern is defined by an outer and inner radius, with which one can either create circles or ring shaped cuts. Rings are particularly useful for fabricating pillars, for which a dedicated circular scan strategy can be employed.

Polygon pattern The main use of polygon patterns when fabricating microstructures is the even removal of the top layer of gold on a lamella (c.f. Fig. 2.14b). Otherwise polygons are rarely used when operating the FEI xT software. The reason for this is that the user does not have control over the generated scan path, which depends strongly on the shape of the polygon. Moreover, occasionally the beam shutter is not fast enough in blanking the beam when skipping across the sample. This then leads to characteristic beam trails dragged across parts of the sample which were not intended to be milled. In these circumstances, it can be helpful to increase the dwell time. However, by circumventing polygon patterns, this issue can avoided all together.

Bitmap pattern FIB software can interpret bitmap files as binary input and generate an according scan path. This is useful for labelling devices or creating complex structures (including black and white photographs). Similar to the polygon pattern, the user does not have control over the scan path. Hence when working with bitmap patterns it is recommended to increase the dwell time to avoid beam blanking artefacts. However, in general, it is advisable to instead use rectangular cuts milled in parallel rather than a bitmap inputs to fabricate transport devices.

The recommended milling method for microstructuring devices is presented in Fig. 2.4c, showcasing part of a four-point structure. Rectangular patterns, roughly as wide as the cut will be deep, are arranged and aligned such that the direction of milling is pointing *towards* the current path in order to reduce redeposition. The T–shaped cut consists of two rectangular bars which overlap minimally. As mentioned above, due to the well-defined scan path, rectangles are strongly preferred over a polygon pattern. The entire device is milled in parallel to minimise redeposition. Subsequently, the side walls of the current path are polished using a CCS at low currents, to create a well-defined rectangular cross section.

The typical consequence of CCS side-wall polishing is a rounded top edge, as sketched in Fig. 2.4d. This is a result of the extended beam profile (in particular when using a plasma FIB), which leads to sputtering at the edge of the sample where material removal is especially effective. This effect can be reduced by using lower beam currents, as their diameter is smaller. However, this simultaneously decreases the sputtering rate, which significantly prolongs the polishing procedure. Alternatively, a protection layer can be deposited on top of the relevant section by means of the gas injection system (GIS), c.f. Sect. 2.1.4. The so-grown structures can typically withstand high beam currents and offer an effective protection of the underlying material.

As mentioned above, minimising the so-called *redeposition* of sputtered material is one of the main challenges during the microfabrication of transport devices. Excessive amounts of redeposition can lead to short circuits or contribute to the parallel resistance measured in any transport experiment.

Generally, redeposition describes the situation when sputtered atoms re-impact on the surface of the sample. The amount and distribution of the redeposition layer depends on the geometry of the milled cut, the scan pattern and on the energy of the incident ions. In the case of a rectangle pattern, as shown in Fig. 2.5a, the layer-by-layer scan path removes redeposited material at the bottom of a shallow trench. As the depth of the trench increases, the redeposition rate rises until it meets the sputtering rate, at which point effective milling is terminated. Due to its distinct scan

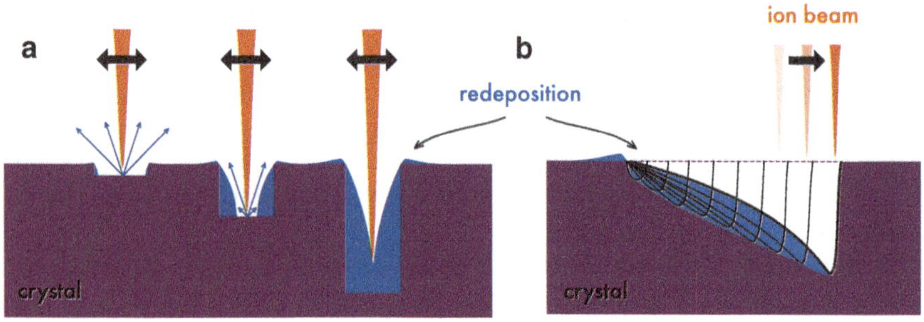

Fig. 2.5 **a** Schematic of a FIB layer-by-layer cut, illustrating the influence of the aspect ratio on the milling efficiency. As the trench grows deeper, the probability of a sputtered atom to escape decreases with decreasing opening angle. Instead, the atoms increasingly impact the sidewalls and form a redeposition layer (in blue). **b** The asymmetric distribution of the redeposition layer of a line-by-cut can be exploited to create clean side walls. I will refer to this process as *polishing*

strategy, the redeposition of a cleaning cross section (CCS) cut is clearly different, c.f. Fig. 2.5b. In this case, the ion beam advances line-by-line in the scanning direction, without reversing to an already milled line. This leads to a build-up of re-deposition on one side of the trench, but minimises the accumulation of re-deposited material in the direction of milling. Therefore, the CCS pattern is ideally suited to fabricate side-walls of microstructures with minimal re-deposition.

2.1.3 Surface Aspects of Ion Beam Fabrication

Invariably, the irradiation of ions will create a damaged layer on the surface of the target material it impinges on. Under certain circumstances, the damage layer can lead to observable consequences in electrical transport measurements, including surface superconductivity induced by gallium doping [9].

A sketch of a sample after ion irradiation is shown in Fig. 2.6a. In general, the effects of the ion beam can be summarised as creating an amorphous damage layer around the crystalline solid. Additionally, if a liquid metal ion source (such as gallium) is used, source ions are implanted into the damage layer, which can locally change the chemical potential of the material. The interaction effects of the ions with a solid is typically well-described with a collision cascade model, which can be simulated using a program called SRIM (Stopping Range of Ions in Matter). SRIM is based on quantum mechanical models for ion interactions which are then simulated using a Monte Carlo method. While SRIM is the most widely used simulation software due to its extensive database, other more advanced codes exist [12]. For instance TRIDYN is a dynamical simulation software package, which takes into account the stoichiometric change of the sample due to ion implantation, atomic mixing and preferential sputtering effects.

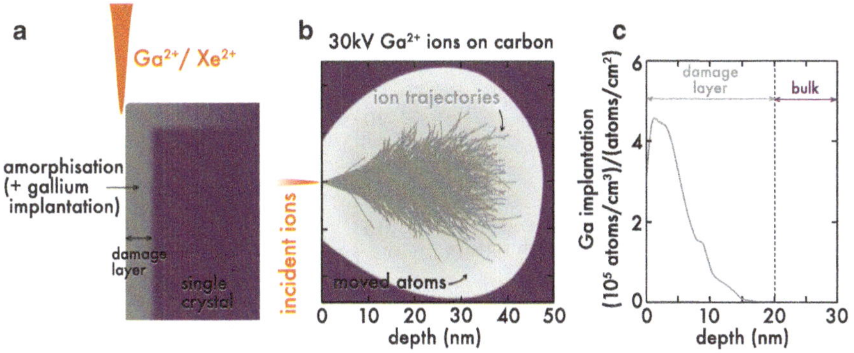

Fig. 2.6 **a** FIB prepared structures are encapsulated by a shell of amorphous material. **b** A sketch of a typical ion collision cascade showing the trajectories of the stopped ions (dark grey) as well as the moved atoms (in light grey). This sketch is based on the simulation of 30 kV Ga^{2+} ions in carbon [10]. **c** In a similar simulation, SRIM was used to find the gallium implantation distribution in $PdCoO_2$ in a situation, where the ions impact the surface at a grazing incidence angle. Figure based on supplementary information of [11]

An illustration of a typical collision cascade is presented in Fig. 2.6b. Here, 30 kV Ga^{2+} ions impinge on the surface of carbon under a normal incidence angle. The resulting ion trajectories, which extend about 20–30 nm into the bulk, are shown in dark grey. The kinetic energy of the ions can further cause atoms to move in an area extending the ionic plume by a further 10–15 nm, which can lead to point defects, for instance. A comparable SRIM simulation was performed for 30 kV Ga^{2+} focused onto $PdCoO_2$. However, in this case, the ions were simulated at a grazing incidence angle, as is the situation during side-wall polishing. The resulting gallium implantation distribution, shown in Fig. 2.6c, follows Gaussian distribution centred around the initial ion collision point, which is approximately 2 nm away from the surface, in this case. The statistical nature of a random walk dictates that implantation of gallium ions is exponentially suppressed by one decade every 6–20 nm, where the ion penetration signal cuts off in panel c [11]. The maximal defect propagation into the bulk of the sample is then another 10–20 nm additionally to the maximum ion penetration, as indicated by light grey plume of moved atoms in panel b.

In summary, the damage layer created by 30 kV Ga^{2+} ions, which are commonly used for microstructuring typically extends 20–30 nm into the bulk. Due to the amorphous nature of the surface layer, it usually adds a negligible resistance contribution and can be ignored for most parts. This assumption, however, does not hold if the resistance of the amorphous damaged layer becomes comparable to the bulk resistance, or even superconducting.

Exactly this situation occurs when compounds of the mono-arsenide family (Nb, Ta), (P, As) are FIB microstructured as a result of a preferential sputtering effect [13, 14]. Figure 2.7a graphically explains the situation using NbAs as an example. While the bulk remains intact, the damage layer is heavily enriched with niobium, leading to superconductivity at low temperatures, as shown in panel b. The reason

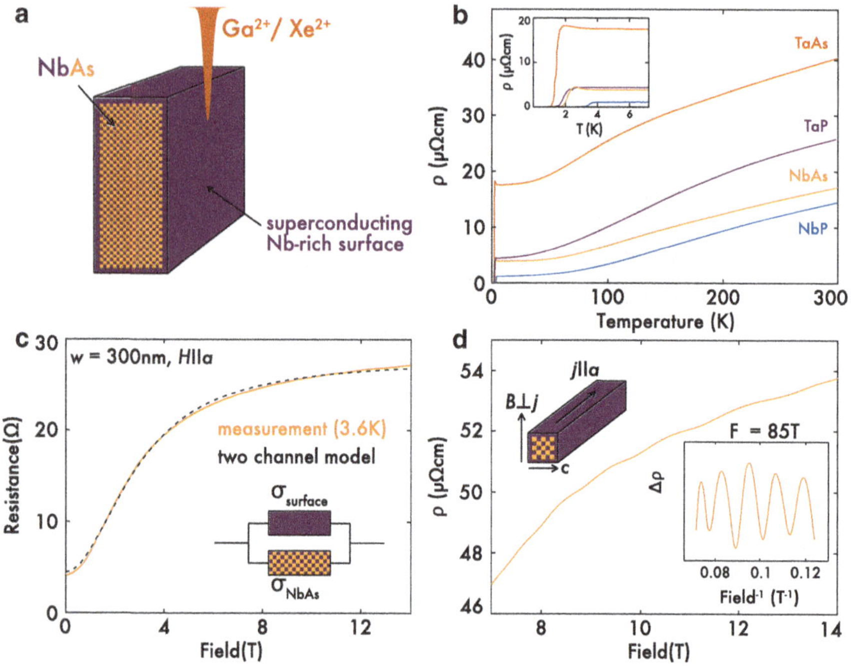

Fig. 2.7 **a** Due to the strong preferential sputtering of arsenic (As) compared to that of niobium (Nb), the amorphous layer of NbAs is predominantly composed of niobium. **b** The resistivity as a function of temperature reveals that FIB milling of the mono- arsenide family (Nb,Ta)(P,As) creates a superconducting shell around the otherwise non-superconducting materials. **c** By fitting a parallel resistor model to the measured magneto-resistance of a 300 nm wide microstructured NbAs four-point bar, a sheet resistance of 5 Ω/sq can be extracted. The current and field configuration are shown in panel d. **d** Quantum oscillations in the same device and field configuration reveal an intact crystalline bulk

for the niobium enrichment in the amorphous layer is the large difference between the sublimation points of arsenic (887 K) and niobium (5017 K). This is confirmed by SRIM simulations for normal incident 60 keV gallium ions, which find a sputter yield of 1.7 per incident ion for Nb and 10.01 for As. Under equilibrium sputtering conditions one may therefore expect a surface composition of around Nb:As ≈6:1.

The same mechanism also leads to a superconducting surface layer with a superconducting transition temperature between T_c ≈1–4 K in TaAs, TaP and NbP, as shown in the inset of Fig. 2.7b. Panel c displays the resistance of a 300 nm wide, microstructured NbAs sample as a function of magnetic field B. Using a simple parallel resistor model, $\sigma_{\text{tot}}(B) = \sigma_{\text{NbAs}}(B) + \sigma_{\text{Nb-layer}}(B)$, the sheet resistance of the amorphous niobium layer can be estimated to be 5 Ω/sq, which is in quantitative agreement with the resistance of FIB-irradiated Nb thin films [15]. For this we assumed the usual quadratic dependence of the magneto-resistance of the NbAs bulk,

$\sigma_{\mathrm{NbAs}} \sim (aB^2 + \text{const.})^{-1}$. Moreover, the amorphous damage layer is approximated as field independent in comparison to the strongly field dependent magneto-resistance of NbaAs, $\sigma_{\text{Nb-layer}} \sim$ const.

Furthermore, although the microstructure is merely 300 nm wide, pronounced Shubnikov–de Haas oscillations (for more detail on quantum oscillations see Sect. 4.2.5) can readily be measured, as presented in panel d. As quantum oscillations are exponentially sensitive to defects, their observation (at the correct bulk value [16]) is a clear indication that the crystalline quality and symmetry are preserved within the bulk of the device.

The method presented above lays out a new route to reliably fabricating superconducting microstructures from appropriate materials. The combination of a high-quality, single crystalline bulk with induced surface superconductivity holds the potential to study proximity-induced superconductivity. This would be particularly interesting for the mono-arsenide metals NbAs and TaAs, as they belong to a class of so-called Weyl semimetals, which exhibit topological properties. If Cooper pairs would be induced in these compounds by proximity, the superconducting state could inherent the non-trivial topological properties of the Weyl metal and form a Weyl superconductor. These are expected to host exotic phenomena, such as finite-momentum Cooper pairs or zero-energy Majorana modes at their surface and are of fundamental interest to improve our understanding of correlated topological systems.

2.1.4 Ion Assisted Chemical Vapour Deposition

A focused beam of ions can not only be utilised to locally sputter material, but also to locally deposit material in a maskless fashion. This direct writing technique operates on the basis of local chemical vapour deposition on the sample surface assisted by the incident energetic ions of the FIB. With this it is possible to deposit amorphous conductive and insulating materials, such as platinum, tungsten, carbon and SiO_2.

The details of the deposition process are sketched in Fig. 2.8a. A small nozzle which is connected to the gas injection system (GIS) is brought into close proximity to the sample surface. Once positioned, it releases a precursor gas into the chamber which partially adsorbs on the surface of the sample. The selected gas determines the type of deposited material and typically consists of metalorganic molecules, which are mostly organic molecules in which a metal atom is embedded (e.g. $(CH_3)_3(CH_3C_5H_4)Pt$ for platinum deposition). After a layer of adsorbed gas has formed, the area is locally irradiated with the ion beam, c.f. bottom panel of Fig. 2.8a. Upon impact with the target material, the energetic ions locally generate so-called secondary electrons with a few eV, which can break the bonds of the molecules on the surface of the sample. The volatile organic components are then pumped out while the heavier platinum atoms bond to the surface. However, the deposited material is amorphous and never pure. Gallium-FIB deposited platinum layers, for instance, have been analysed and found to contain 46% Pt, 24% C, 28% Ga, and 2% O [17], although the exact composition varies depending on the local growth conditions (gas

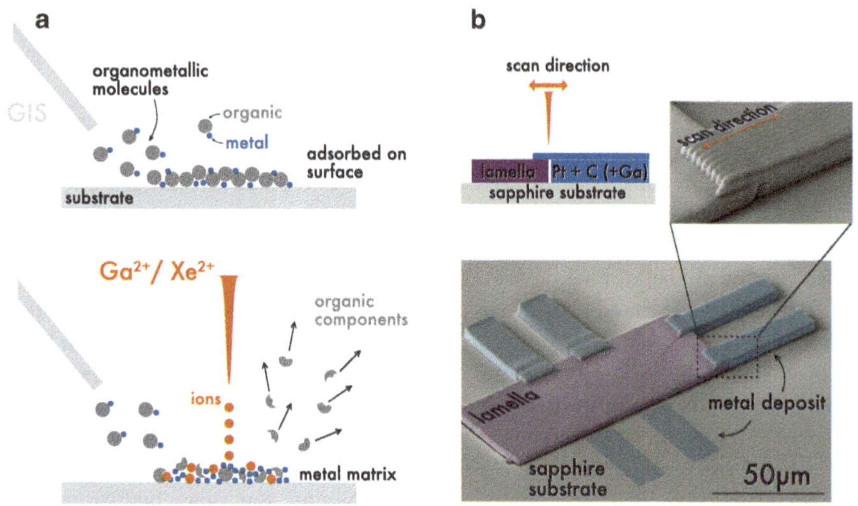

Fig. 2.8 **a** Schematic image of the FIB assisted deposition process. Top: A gas of organometallic molecules is inserted by the gas injection system (GIS) and adsorbed on the sample surface. Bottom: The ion bombardment causes the organic components to crack, leaving behind a conductive matrix. **b** SEM image of metallic leads deposited around a lamella. The deposition is performed in two steps in order to minimise self-shadowing effects

flow, irradiation density, angle of irradiation, ...). Moreover, the resistivity of the deposited metallic layers is typically several orders of magnitude higher than that of pure metals.

Due to their comparatively high resistivity, FIB deposited bonds are not ideal for contacting microstructured samples. And although deposited tungsten becomes superconducting below $\sim$8 K, the sample contact resistance may still remain high. However, the main advantage of FIB deposited contacts is their strength and chemical resistance. Due to the formation under ion bombardment, the Pt strips are firmly anchored to the substrate and can sustain extreme conditions, such as hydrostatic pressure environments or rapid thermal changes. For these situations it is recommended to use a combination of gold sputter deposition, which creates low ohmic contacts with platinum bonding to secure and protect them.

When contacting microstructures (lamellae or crystal platelets), the recommended growth strategy is to first create a base layer, leaving a $\sim$1 μm gap to the sample in order to avoid self-shadowing effects, c.f. Fig. 2.8b. Once the base is approximately the same height as the sample, they can be connected by extending the area of irradiation to overlap with the sample. Note how the ‘growth lines’, which indicated the scan direction are oriented parallel to the future direction of current flow, in order to reduce with resistivity along that direction. Further details regarding the growth process are described in [18].

Finding the optimal deposition parameters is not always easy and typically requires some testing. Unfortunately, there is no universal recipe which can be applied, as the growth rate is sensitive to the local environment, such as the gas flow rate and dispersion (which should ideally be laminar), the ion energy and dose, the dwell time, and the refresh time. It can even happen that nominally identical settings will work on parts of the device, but not on others due to local changes in the gas distribution. Recommended settings are to use a beam overlap of −50%, 50 ns dwell time and 1 µs refresh time. A conservative estimate for the ion beam current I in pA is given by I [pA] $\approx 5 \cdot A$ [µm^2], where A is the area of irradiation in µm^2. For instance, the platinum structures shown in Fig. 2.8b each have an area of approximately 15 µs × 50 µm = 750 µm^2, hence the empirical formula suggests using an ion current of 3.75 nA. In practice, the current was initially set to 2.5 nA and then increased to 9 nA after a thin layer was successfully deposited.

With this, we have discussed the relevant functions of the FIB necessary for microstructuring in detail. The next section will illustrate how these can be used to create high-quality transport devices from quantum materials.

2.2 Transport Device Fabrication

The FIB is an ideal instrument to fabricate transport devices out of metallic materials in order to investigate their electronic properties. In particular, a FIB based fabrication approach offers the opportunity to design the shape and dimensions of a sample around the physical property or phenomenon of interest. For instance, by fabricating long meander structures from high conductivity metals, the overall resistance of the device can be enhanced by over an order of magnitude, which facilitates transport measurements requiring small excitation currents for temperatures in the millikelvin range [19]. Furthermore, by reducing the cross section of a conductor to sub-µm^2, critical current density measurements have become feasible in materials in which achieving the critical current threshold would otherwise be impracticable [20]. Moreover, devices with triangular cross sections have been fabricated to study topological surface effects [21] and possible hydrodynamic electron flow was investigated in channel geometries [11, 22]. For standard electrical transport characterisation measurements a typical device design is fabricated in which several daisy-chained 4-point transport bars are oriented along different principal crystallographic directions, c.f. Fig. 2.15. These devices are termed 'U'-structures, because their shape resembles the letter 'U'.

In the following section I will first present the process flow for fabricating a standard 'U'-structure from a lamella, followed by a detailed description of the procedure developed to create ballistic devices from delafossite crystals. This is intended as a step-by-step guide detailing the microfabrication process.

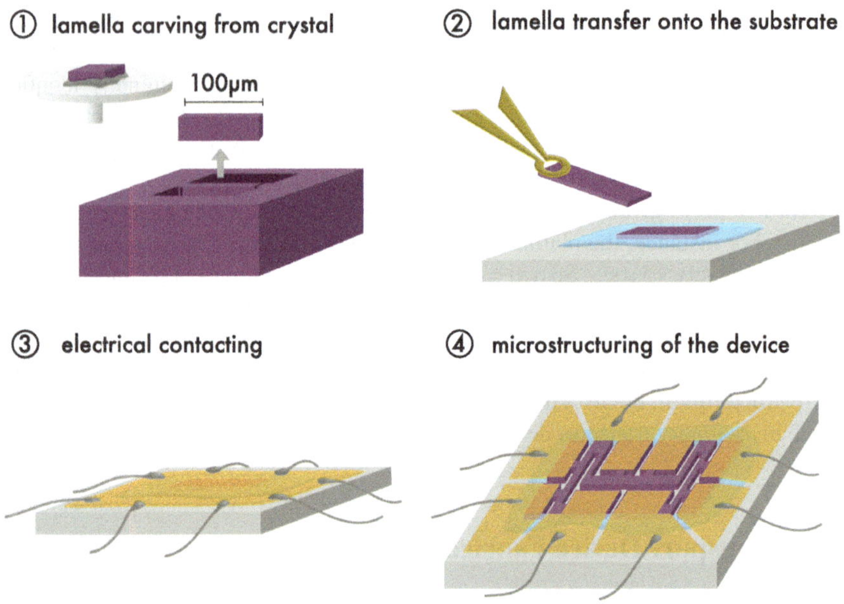

Fig. 2.9 Process flow of the fabrication of microstructured devices. ① After the crystal is firmly mounted on an SEM stub (top left), the first step towards making a microstructured device is to FIB carve a lamella from a single crystal. ② Next the lamella is transferred ex-situ into a thin layer of epoxy (light blue) on a sapphire substrate (grey). ③ After the epoxy has cured, a thin layer of Au (yellow) is sputtered onto the substrate, creating a top contact to the lamella. The gold is further connected to silver wires using Ag epoxy. ④ In the final step, the lamella (purple) is patterned into a transport measurement using the FIB to define a current path through the conductor

2.2.1 Lamella Sample Preparation

Single crystals of quantum materials commonly grow as small, often irregularly shaped crystallites, in particular when synthesised for the first time. Above a critical crystallite size ($\gtrsim$100 μm), the electrical connections needed to perform a transport experiment (i.e. current and voltage wire pairs) can directly be attached by hand using silver epoxy. However, if the sample has a maximum linear dimension shorter than ~100 μm, the widths of the contacting wires themselves become problematic, even for the smallest commonly available wire diameter of 10 μm. Moreover, on these length scales difficulties in achieving a homogenous current flow through the conductor may arise. The situation is particularly challenging if one is interested in the out-of-plane transport of an anisotropic material in which the crystallites only grow in the form of thin platelets. Faced with exactly this issue, P. Moll developed a FIB based method in which a small slice of a crystal, a so-called *lamella*, is extracted and subsequently microstructured in order to create a microscopic transport device, which accesses both the in-plane and out-of-plane transport direction [23]. The main steps following Moll's method are graphically summarised in Fig. 2.9.

① To begin, a lamella (typically $\approx 100 \times 50 \times 3\,\mu m^3$, but smaller if necessary) is excavated from an oriented single crystal (shown in purple) with the FIB. ② Subsequently, the lamella is transferred onto an insulating substrate (shown in grey) and fixed using an epoxy resin (light blue). The transfer process is facilitated using a 3-axis hydraulic micromanipulator, to which a kapton tip is attached (shown in yellow). The lamella adheres to the tip electrostatically and can then be placed precisely into the epoxy. ③ After the epoxy is cured, a layer (100 nm) of gold is sputtered on top of the lamella to form a low-ohmic electrical contact. Silver wires are subsequently attached to the gold using silver epoxy, enabling further electrical connections to be easily made by soldering. ④ In the final step, the lamella is loaded into the FIB once more to define and structure the desired transport device and separate the gold layer into individual contacts.

This method can be used to process most conductive materials and with some practice the device fabrication process can be completed in less than two days. However, there are many fabrication details which one should be aware of and pay attention to in order to create a correctly functioning device, especially when intending to perform transport measurements. For this reason I will go over steps ①–④ of Fig. 2.15 again, with the intention of describing all details of fabrication. Readers who are not interested in the specifics of device fabrication may wish to proceed the next chapter starting in Sect. 3.1.

1. Lamella carving from a single crystal

To begin, the crystal must be firmly mounted onto and electrically connected to an SEM stub, c.f. Fig. 2.9, ①. This is best achieved with conductive SEM silver paint, for instance *Silver Conductive Adhesive 503, #12686-15* from EMS, which is easily mechanically removable if required. This task serves the purpose of grounding the crystal electrically and thermally, and prevents spatial drifts.

Ideally at this stage the crystal is mounted such that one crystallographic axis of interest is directed perpendicular to the SEM stub. The FIB does allow pre-tilting of the sample, but with significant misalignment one can run into tilt angle limitations of the stage. With the oriented crystal mounted firmly in place, the lamella fabrication process can commence. A series of images documenting this process is displayed in Fig. 2.10. In this example, the gallium FIB was used to fabricate a lamella, however in general, this process is more time efficient utilising the xenon FIB. The process flow using these two systems is identical, but the extended beam tail of the xenon FIB must be kept in mind and accounted for when polishing the side walls of the lamellae.

(a) This specific crystal was mounted with one axis of interest directed perpendicular to the SEM stub, such that the trenches defining the lamella are cut at a tilt of 52° with respect to the SEM beam, i.e. parallel to the ion beam column. The trenches are cut using two rectangular cross section patterns (RCS). Their spacing depends on the selected beam current, which determines the beam diameter and hence the additional gap which should be left between the trenches. In order to FIB-polish the side walls of the lamella in a later step, 2 μm should

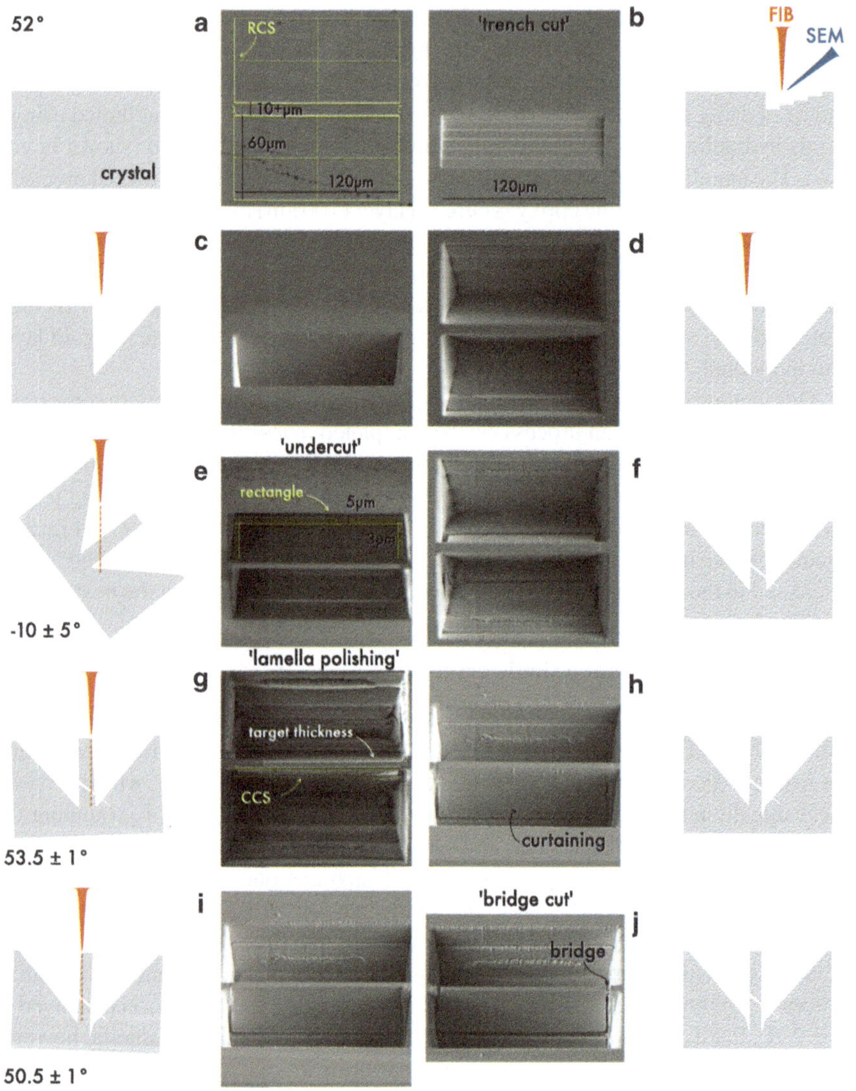

Fig. 2.10 The lamella excavating process showcased on a single crystal of $LaIrIn_5$. The picture series lays out the how initially a *trench* is cut (**a**–**d**) followed by the *undercut* of the lamella (**e**–**f**). Subsequently the front and backside of the lamella are polished using the FIB beam (**g**–**i**) and lastly only a thin bridge is left behind which holds the lamella in the crystal (**j**). The yellow lines indicate the particular scan strategy (regular cross section (RCS), rectangular cut, or cleaning cross section (CCS), which was used for the particular pattern

be additionally included for both sides of the lamella. Hence, the optimal gap between the trenches sums up to: lamella target width $+ 2 \times 2\,\mu$m for polishing $+2\times$ beam radius. In this example a $10\,\mu$m wide gap was chosen for a 21 nA beam of gallium ions at 30 kV to create a 2–3 μm wide lamella. Further, the desired depth ('*z size*') of the trench will dictate the optimal width of the trench (60 μm in this case) in order to achieve an efficient undercut, c.f. panel e.

(b) After some time the partially milled staircase of the RCS pattern is clearly visible in the crystal. The RCS milling strategy is optimised for fast removal of large volumes and leads to the characteristic triangular trench cross section designed for lamella cutting.

(c) The first RCS is finished. The light grey halo around the trench are signs of redeposited material.

(d) The second RCS is completed. From the top down view it is clear the side walls of the lamella are not perfectly parallel to each other but come together to form a slightly trapezoidal cross section.

(e) The next step is the *undercut*, which defines the length and width of the lamella. For this, the stage is tilted from $+52° \rightarrow -10 \pm 5°$, such that three rectangular cuts can be placed on the face of the lamella. A bridge on either side at the top of the lamella is left connected to the crystal for stability and to dissipate heat.

(f) The undercut has milled through the entire depth of the lamella, as indicated by the U-shaped cut in the lower side wall of the trench. Importantly, the undercut was performed *before* the polishing steps to avoid redeposition on the sides of the lamella.

(g) When polishing the sides of the lamella, the crystal is tilted back to $+52°$, such that the surface is parallel to the ion column. In order to remove the trapezoidal cross section, the sample is in-fact over-tilted by 1–2°. This so-called *slope angle* is a characteristic feature of ion milling and depends on the target material as well as the energies of the ions. For the material and beam conditions in this example, 1.5° additional tilt angle was added (i.e. the stage is tilted to $+53.5°$). A cleaning cross section (CCS) cut is then used to polish the angled side wall. It can be helpful to place a marker indicating the target thickness of the lamella (typically around 2–3 μm), in order to avoid over-polishing.

(h) A common occurrence when polishing with the FIB is the appearance of so-called *curtaining*, which describes irregular and 'streaky' side surfaces after ion irradiation. Curtaining is the result of the spatial variation of the sample sputtering rate due to rough or uneven top surfaces, which deflect the ion beam. Additionally, this effect can be amplified in the event of ion channeling by the crystalline solid. Curtaining can be reduced by using a CCS and gradually (and patiently) polishing the side wall of the lamella from the bottom towards the top. Moreover, altering the beam current or dwell time can aid the removal of curtaining. In some cases, it may be helpful to eliminate the roughness of the top surface completely by milling it away in the same manner as an undercut, c.f. panel e. A further strategy to minimise curtaining is to deposit a protective layer (e.g. from platinum) on top of the crystal (i.e. prior to panel a) before milling the initial trench, c.f. Fig. 2.4d. Additionally, the dense top layer shields the top edge of lamella from beam tail rounding effects, in particular when using the xenon FIB.

(i) Once polished from both sides, the surfaces of the lamella should look entirely even. Any horizontal feature running across the side of the lamella indicates that the face is not completely flat yet and requires further polishing.

(j) In the final step, the connections of the lamella to the crystal are reduced to one small bridge, less than 1 μm^2 in cross section. The smaller the bridge, the easier it will be to break when the lamella is transferred.

Table 2.1 Ion beam current guideline for lamella fabrication. Unless otherwise specified an acceleration voltage of 30 kV was used and a dwell time of 1 μs for milling. These parameters have proven to work well on a wide range of materials. However, as every compound behaves differently under ion irradiation, the values quoted above are intended to be a starting point when working on new materials and should then be adapted to achieve maximal efficiency. The typical duration of the process steps are quoted for a standard sized lamella with comparable milling properties to silicon

	GFIB	PFIB	Typical duration
Trench cut	21–65 nA	60–200 nA	2–4 h
Undercut	21 nA	15–60 nA	15–30 min
Lamella polishing	9–21 nA	15–60 nA	30–60 min per side
Bridge cut	2.5 nA	4 nA	5 min
Lamella gold removal	8 kV, 5 nA	4 nA	2 min
Structuring	2.5 nA	4–15 nA	1–3 h
Structure polishing	0.8 nA	4 nA	1 h
Gold contact cutting	21 nA.	15 nA	30 min
			Total: 6–11 h

Table 2.1 summarises the typical beam currents used during the lamella fabrication process and subsequent structuring when using a gallium FIB or a xenon plasma FIB. These parameters have proven to work well with numerous materials. However, as the interaction of the ion beam with the sample varies depending on the ion source as well as target material, these values are only intended as a first reference point when working with new materials. In any case the beam currents should be optimised for every compound individually. Note that the currents can vary greatly depending on the ion source (gallium or xenon). This is due to the changing beam diameter and beam tail as a function of current, as discussed in Fig. 2.3.

In general, the most time efficient way of fabricating a microstructured device is to exploit the high sputtering rates of the xenon plasma FIB when excavating and polishing the lamella and taking advantage of the superior performance of the gallium FIB at low currents when structuring the transport device.

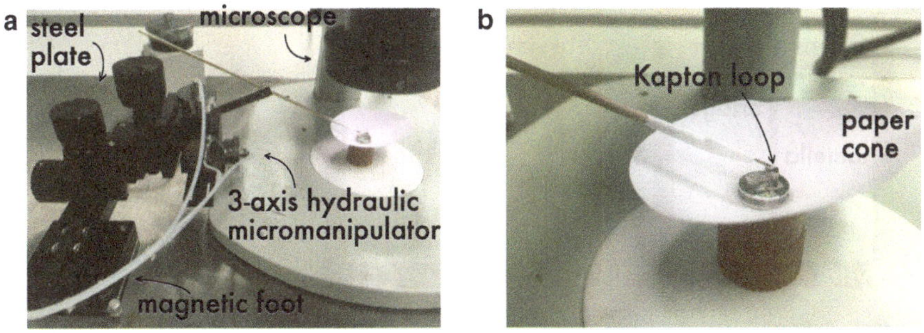

Fig. 2.11 **a** Photograph of a transfer setup. A stainless steel plate on top of a firm table create a robust, low vibration foundation. The 3-axis hydraulic micromanipulator possesses a magnetic base which is fixed to the steel plate. **b** A Kapton loop is attached to a wooden skewer using teflon tape. A clean paper cone envelopes the aluminium SEM stub in order to capture unintentionally launched lamellae

2. Lamella transfer onto the substrate

After the lamella is polished and the bridge is cut, it is transferred onto an insulating substrate (typically sapphire or Si/SiO_2). This may require some practice and patience to achieve optimal results. This undertaking can be divided into three main tasks

- breaking off the lamella and positioning it on the Kapton tip (without losing it),
- painting a layer of epoxy onto the substrate which is approximately the same thickness as the lamella, and
- transferring the lamella into the epoxy.

A typical setup for transferring lamellae is shown in Fig. 2.11. I recommend crafting a paper cone around the SEM stub on which the crystal is mounted and performing the transfer process within the cone. This greatly reduces the time to retrieve a lamella should it unintentionally be lost.

To facilitate the transfer process we utilise a three-axis hydraulic micromanipulator (*Narishige Model MHW-3*), which can be magnetically fixed to a stainless steel plate which is placed on a steady table. This forms the foundation for a low-vibration environment for precise transferring. Next, a long wooden skewer to which a small Kapton tip is attached, is mounted into the micromanipulator. These Kapton loops (*M2-L18SP-10*) were originally designed for mounting X-ray samples, but are ideally suited for handling lamellae by exploiting electrostatic forces.

With these tools the lamella in the crystal (c.f. Fig. 2.12a) can then be approached and gently broken off. Depending on the cross section of the bridge, slight force may be necessary to fracture it. This can be achieved by carefully pushing against the lamella with the Kapton tip (c.f. Fig. 2.12b). Once the lamella is detached from the crystal, it should be placed on the loop such that the lamella forms the lowest point, as shown in Fig. 2.12c. Subsequently, a thin layer of epoxy is spread in the centre of the substrate (we use the two-component *Araldite® rapid* epoxy). For this task I recommend using a single haired brush, which can be made, for instance, by attaching an eyelash to a skewer. Alternatively, a toothpick can be shaved down until

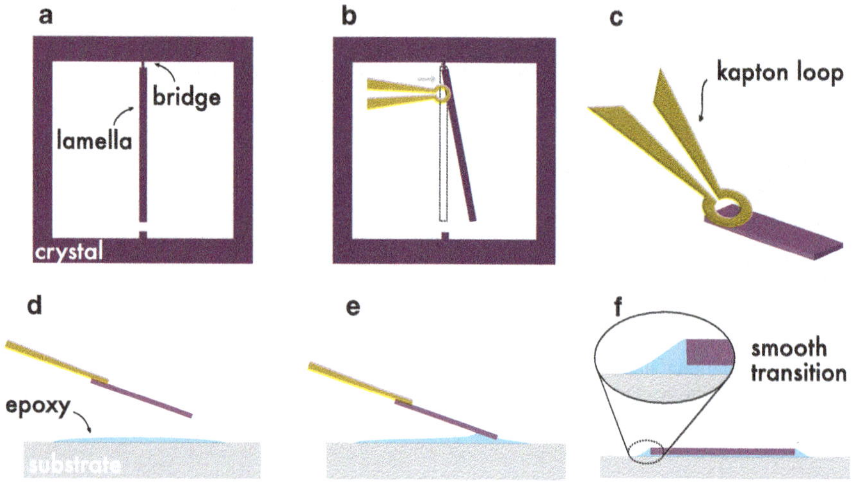

Fig. 2.12 The lamella transferring process. **a** The finished lamella from Fig. 2.10j displayed from a top-down view. **b** Using the tip of a Kapton loop the bridge of the lamella is broken and the lamella should ideally electrostatically stay on the tip. **c** For a successful transfer, the placement of the lamella on the loop should such that the lamella is the lowest point. **d** After an epoxy droplet roughly equal to the thickness of the lamella is applied to the substrate, the lamella can be gently lowered into the epoxy layer **e**. **f** Importantly, the epoxy should form a smooth transition from the sapphire substrate to the top of the lamella, such that a ramp is formed for the gold layer deposition in the next step

only a single fibre remains at its end. Finally, the lamella is slowly lowered into the epoxy using the micromanipulator. Once the lamella comes in contact with the epoxy (c.f. Fig. 2.12e), the surface tension will draw the lamella onto the epoxy and self-align the surface of the lamella parallel to the substrate, while forming a smooth transition from the substrate to the top of the lamella, c.f. Fig. 2.12f. Lastly, the epoxy is cured at 100 °C for 1 h to remove any remaining water.

3. Electrical contacting

Electrical connection to the lamella is established by sputtering a thin layer of gold across the entire device. For this purpose a shadow mask is created from Kapton tape, which covers the edges of the substrate, such that only the centre of the substrate is sputtered with gold, c.f. orange area in Fig. 2.13a. In an integrated sputter chamber system, the surface of the lamella is first cleaned using low energy argon etching and subsequently contacted with a 5 nm sticking layer of titanium followed by a 150 nm layer of gold. Subsequently, silver wires are attached to the sapphire substrate using a two-component silver epoxy (*Epotek® EE129-4*), which is cured at 100 °C for 1 h. This step is intentionally performed before the lamella is further processed in order to avoid heating up the microstructured device, which could potentially cause internal strain leading to damage within the final device, whose minimum lateral feature size may be microns or less.

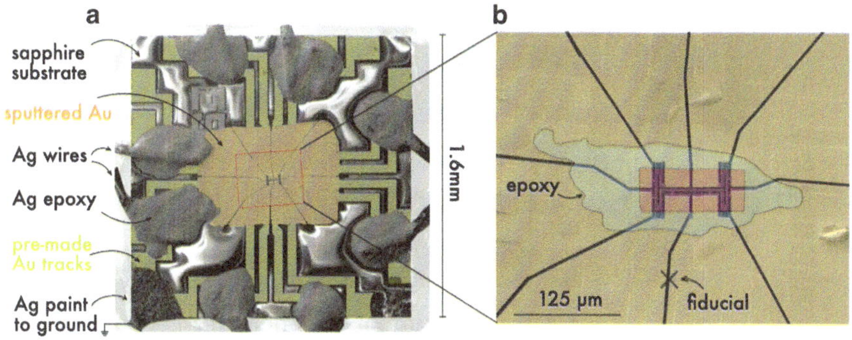

Fig. 2.13 **a** A finished microstructured device shows how the lamella (in the centre) is electrically contacted. The sputtered gold layer (orange) contacts the top surface of the lamella to the pre-made gold tracks (yellow) on the sapphire substrate. These are then connected to silver wires using silver epoxy. **b** A closeup of a finished lamella shows how the gold layer connects to the top surface of the lamella. The fiducial cross was used for drift correction during the structuring of the lamella

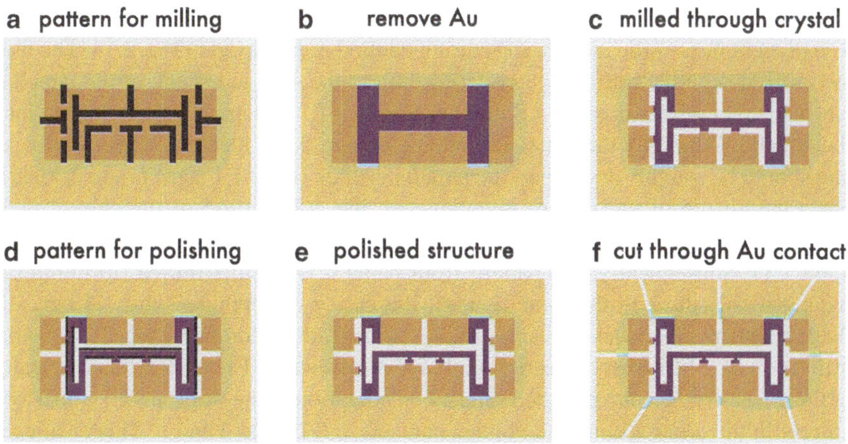

Fig. 2.14 Process flow of the lamella microstructuring procedure. **a** The first step in structuring a lamella is to design the FIB cuts (in black) based on the dimensions of the lamella and the intended functionality of the device. **b** The top layer of Au is removed using a polygon pattern in the places where current will be flowing, to avoid short circuiting. **c** The design of panel a is milled into the lamella. **d+e** Cleaning cross sections are used to polish the sides of structure and simultaneously narrow down the voltage contacts. **f** Finally the top gold layer is removed in certain areas to create separated current and voltage contacts to the lamella

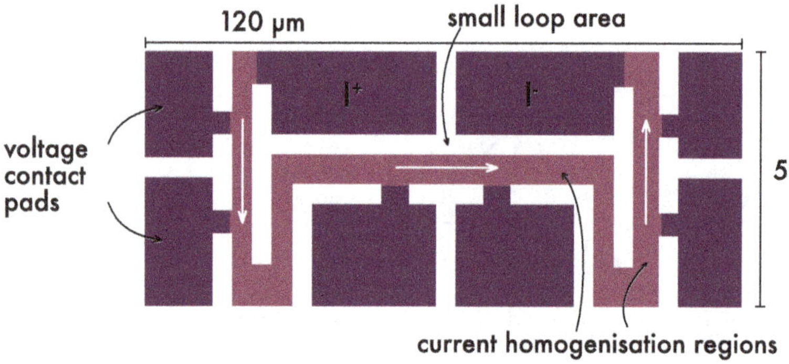

Fig. 2.15 Starting from a rectangular lamella, a 'U'-shaped sample design offers an optimised layout to determine the resistivity anisotropy of the lamella, including one redundant voltage pair along the vertical axis. The white arrows indicate the path of the current

4. Microstructuring the device

The process flow of the lamella structuring method is displayed in Fig. 2.14. The first step is to design the milling pattern according to the size and shape of the lamella, c.f. black rectangles in panel a. In this example, a 'U-structure' will be milled into the lamella. Generally the pattern should be composed of rectangle cuts (not lines) and should be roughly as wide as the lamella is thick (i.e. a 1:1 aspect ratio is ideal). Once the design of the pattern is completed, the gold on top of the current path is removed using a polygon pattern (c.f. panel b and beam current parameters in Table 2.1). Subsequently, the pattern from panel a is parallel-milled (to minimise redeposition!) through the lamella, until the epoxy or sapphire substrate beneath the lamella are clearly visible, c.f. panel c. If the milling duration exceeds ~30 min utilisation of a FIB drift correction tool is recommended. Thereafter, the side walls of the microstructure are polished (under consideration of the slope angle) with a low current in the areas around the voltage contacts, c.f. panels d and e. This ensures a clean finish, such that the size of the conductor and hence the resistivity can accurately be determined. Lastly, the gold layer around the lamella is separated in order to create individual electrical contacts to the lamella, c.f. panel f. Finally, the SEM imaging tools can be used to directly determine the length l, width w and thickness t of the 4-point bars, which are needed when converting the measured resistance R into the resistivity $\rho = \frac{w \cdot h}{l} R$.

The benefits of the 'U-structure' design are highlighted in Fig. 2.15. The area of the lamella is optimally divided to accommodate three daisy-chained 4-point bars, yet leaving space for the current to homogenise. Moreover, the enclosed loop area is small, such that these structures are ideally suited for pulsed field measurements.

An alternative method for mounting a lamella on a sapphire substrate which relies on FIB deposited platinum bonds (in blue) is displayed in Fig. 2.16. This strategy is employed in situations where epoxy can not be used, for instance, when fabricating samples for transport measurements under applied hydrostatic pressure. For these

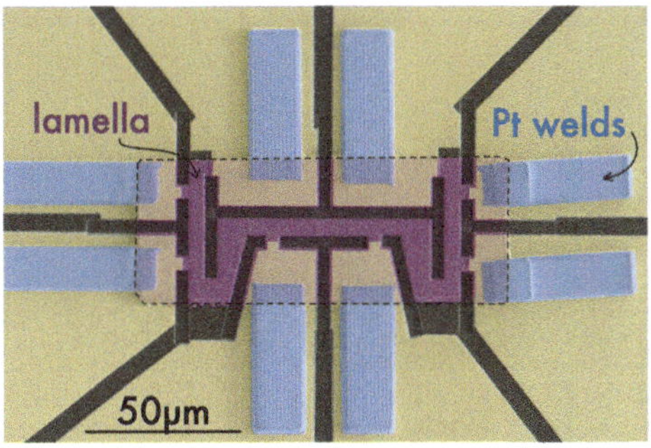

Fig. 2.16 Epoxy-free mounted sample. Alternatively to using epoxy for fixing the lamella onto a substrate, FIB deposited platinum bonds (in blue) can be used to make electrical connections and hold the lamella into place. This method can be beneficial in situations where epoxy can not be used for sample mounting, for instance for transport measurements under hydrostatic pressure conditions utilising piston cylinder or diamond anvil cells

experiments, the samples are submersed in a liquid pressure transmitting medium (e.g. oil or alcohol mixture) and subsequently pressurised. Under these conditions, the comparatively soft epoxy starts to flow, which can fracture the contacts of the sample.

To this end, the lamella is not transferred into epoxy during the fabrication process, but is instead placed on the bare sapphire substrate, to which it adheres electrostatically. It is then coated with a layer of gold (∼100 nm) for the first time. This is necessary to ground the lamella during the subsequent deposition of the platinum bonds. Details of the platinum deposition process are given in Fig. 2.8. Thereafter, a second layer of gold is deposited onto the device to reduce its two-point round-trip resistance to less than 5 Ω. Lastly, the lamella is structured as described in Fig. 2.14. In this fashion structures have readily been measured in piston cylinder cells in pressures up to 2 GPa without contact failure. However, as the lamella electrostatically adheres to the substrate, it is possible that the pressure transmitting medium can not flow evenly beneath the lamella. Hence, further development to improve the hydrostaticity of the measurements is required.

2.2.2 Ballistic Device Preparation from Delafossite Crystals

When interested in the in-plane transport properties of a crystal that has grown naturally in the shape of a thin platelet ($\lesssim$20 µm), the tedious process of lamella fabrication can be bypassed by directly transferring the crystal platelet onto a substrate. This approach is used to fabricate ballistic devices from the delafossite metal $PdCoO_2$. The specific microstructure presented in Fig. 2.17 is designed around investigating transverse electron focusing (TEF), described in detail in Sect. 3.4. A key feature of TEF devices are the narrow (250 nm) and closely spaced constrictions (1 µm), c.f.

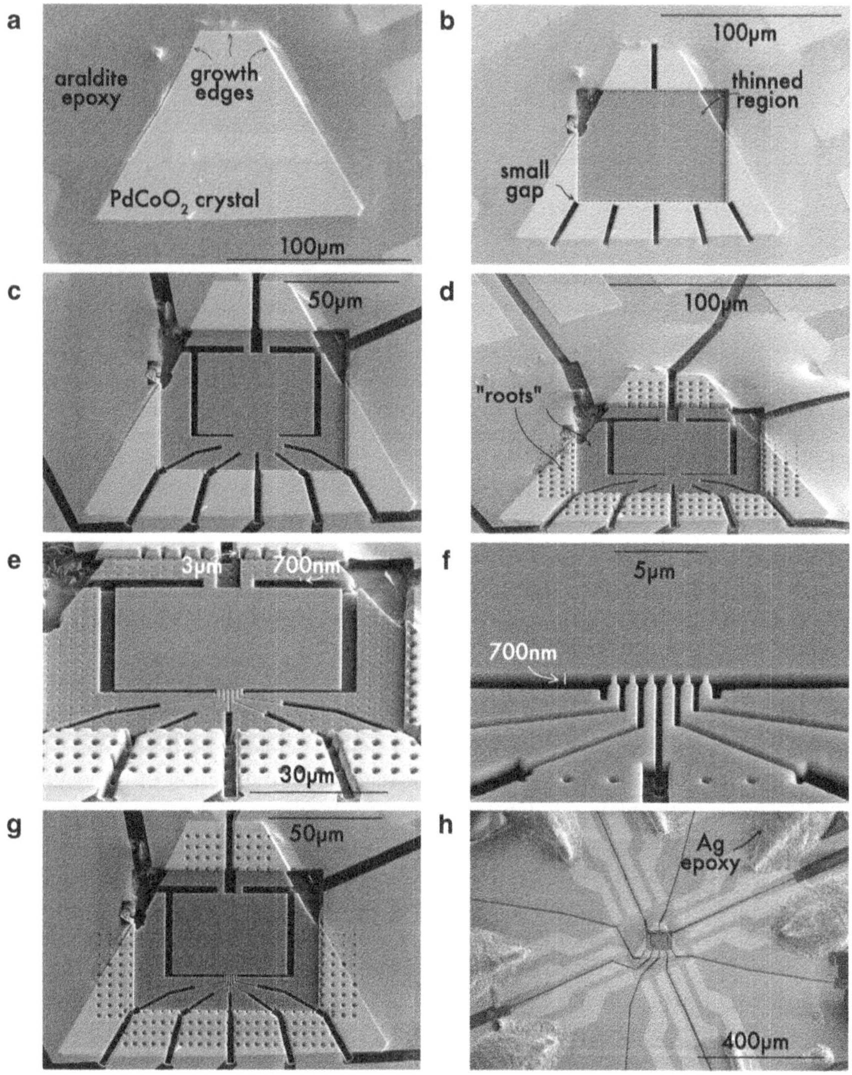

Fig. 2.17 FIB fabrication strategy of a platelet crystal ($\lesssim$20 μm in thickness). Instead of fabricating a lamella, the crystal can directly be mounted onto a substrate and top contacted. The structure presented here is designed to measure transverse electron focusing (TEF), which I discuss in Sect. 3.4

panel f, which are fabricated by milling narrow trenches into the crystal. For this, the thickness t of the crystal is ideally as wide as the trenches, to achieve a clean FIB cut and minimise the risk of short-circuits due to redeposition. To this end, the crystal platelet is thinned down in its centre to less than 1 μm. In the following I will present fabrication details of such a device, following panels a–h of Fig. 2.17. This device is exclusively fabricated using the gallium FIB, although the cuts of panel b can also be milled using a xenon plasma FIB.

(a) Care is taken to select a crystal with as few step edges as possible and with well-defined growth edges, such that the crystallographic orientation can be directly determined (in $PdCoO_2$). The crystal platelet is then transferred into a thin epoxy droplet using the 3-axis micromanipulator shown in Fig. 2.11. A thin epoxy layer is essential to ensure that the crystal surface is perfectly parallel to the substrate. After the epoxy is cured, the sample is masked, contacted with a sputtered gold layer and connected with silver wires in exactly the same fashion as the lamellae described above.

(b) Next, a large region in the centre of the crystal is thinned down to below 1 μm thickness. This is done using a current of 65 nA in the GFIB using a rectangle pattern and the 'dynamic all directions' scan option. Subsequently, $\sim$10 μm wide rectangular cuts are made through the remaining (thick) parts of the crystal in order to define current and voltage leads, again at a current of 65 nA. A small gap is left to avoid redeposition in the central area.

(c) The rectangular cuts in the central region of the crystal were milled with 2.5–9.3 nA and define the 'active' measurement region. The sides of the central rectangle are polished with 2.5 nA under an angle of $+1^\circ$ degree with respect to the normal milling direction to correct for the slope angle and obtain flat and well-defined side walls.

(d) A concern in strongly anisotropic electronic systems is the homogenous flow of current through the conductor. There are multiple strategies addressing this challenge including, for instance, the commonly used long, current homogenising meanders. However due to the finite size of the crystal and the highly multi-contact nature of the TEF devices, this approach is not feasible. Instead, we decided to pattern holes through the entire depth of the crystal with 47 nA and 2 ms dwell time, in order to promote a homogenous current flow between the palladium layers. Similar 'roots' were milled into the central area of the device using an ion beam current of 2.5 nA and 2 ms dwell time. The idea of this procedure is that the amorphous FIB-damage layer and redeposition in the roots increases the interlayer coupling between the highly conducting palladium layers and therefore homogenises the current over a shorter distance than in absence of the roots. Although we do not see signs of inhomogeneous current flow, we have not yet performed a quantitative measurement which determines the effectiveness of the roots.

(e) The channels leading up to the narrow constrictions are patterned at a current of 80 pA. Milling long and thin channels is recommendable, as they act as flexures and reduce the risk of mechanical failure (cracking) of the thin constrictions due to differential thermal contraction-induced strain at low temperatures.

(f) The constrictions are milled using an array of cleaning cross sections (CCS) and a current of 40 pA, while correcting for the slope angle of $+1^\circ$.

(g) The final device. If the constrictions are thinner than 350 nm, a layer of Araldite® rapid epoxy is added on top of the finished device (not shown here) to avoid fracturing of the constrictions due to differential thermal contraction induced substrate strain.

(h) An overview image of the sapphire chip displays the delafossite crystal in the centre, connected by a top layer of gold and silver epoxy contacts to silver wires.

This concludes the description of the microstructure device making process.

References

1. Yan C, Kumar S, Pepper M, See P, Farrer I, Ritchie D, Griffiths J, Jones G (2017) Temperature dependence of spin-split peaks in transverse electron focusing. Nanoscale Res Lett 12:10–15
2. Smith NS, Notte JA, Steele AV (2018) Advances in source technology for focused ion beam instruments. Mater Res Bull 39(4):329–335
3. Smith NS, Skoczylas WP, Kellog SM, Kinion DE, Tesch PP, Sutherland O, Aanesland A, Boswell RW (2006) High brightness inductively coupled plasma source for high current focused ion beam applications. J Vac Sci Technol B 24:2902
4. Kawasaki S (1967) The diffusion of xenon in silver. J Nucl Sci Technol 4:239–243
5. Volkert CA, Minor AM (2007) Focused ion beam microscopy and micromachining. Mater Res Bull 32:389–399
6. Torre R, Larkin J, Singer A, Meller A (2012) Fabrication and characterization of solid-state nanopore arrays for high-throughput DNA sequencing. Nanotechnology 23:385308
7. Fu XL, Li PG, Jin AZ, Chen LM, Yang HF, Li LH, Tang WH, Cui Z (2005) Gas-assisted etching of niobium with focused ion beam. Microelectron Eng 78:29–33
8. Yoon H, Kim C, Lee H, Ahn S (2017) Advanced scanning paths for focused ion beam milling. Vaccum 143:40–49
9. Hermannsdörfer T, Heera V, Ignatchik O, Uhlarz M, Mücklich A, Posselt M, Reuther H, Schmidt B, Heinig K-H, Skorupa W, Voelskow M, Wündisch C, Strotzki R, Helm M, Wosnitza J (2009) Superconducting state in a gallium-doped germanium layer at low temperatures. Phys Rev Lett 102:217003
10. Wu J, Bollinger AT, He X, Božović I (2017) Spontaneous breaking of rotational symmetry in copper oxide superconductors. Nature 547:432–435
11. Moll P, Kushwaha PJW, Nandi N, Schmidt B, Mackenzie AP (2016) Evidence for hydrodynamic electron flow in $PdCoO_2$. Science 351(6277):1061–1064
12. Hofsäss K, Zhang H, Mutzke A (2014) Simulation of ion beam sputtering with SDTrimSP, TRIDYN and SRIM. Appl Surf Sci 310:134–141
13. Bachmann MD, Nair N, Flicker F, Ilan R, Meng T, Ghimire NJ, Bauer ED, Ronning F, Analytis JG, Moll PJW (2017) Inducing superconductivity in Weyl semimetal microstructures by selective ion sputtering. Sci Adv 3(5)
14. Ramshaw BJ, Modic KA, Shekhter A, Zhang Y, Kim E, Moll PJW, Bachmann MD, Chan MK, Betts JB, Balakirev F, Migliori A, Ghimire NJ, Bauer ED, Ronning F, McDonald RD (2018) Quantum limit transport and destruction of the Weyl nodes in TaAs. Nat Commun 9:2217

15. de Leo N, Fretto M, Lacquaniti V, Cassiago C, Ortenzi LD, Boarino L, Maggi S (2016) Thickness modulated niobium nanoconstrictions by focused ion beam and anodization. IEEE Trans Appl Supercond 26(3)
16. Luo Y, Ghimire NJ, Wartenbe M, Choi H, Neupane M, McDonald RD, Bauer ED, Zhu J, Thompson JD, Ronning F (2015) Electron-hole compensation effect between topologically trivial electrons and nontrivial holes in NbAs. Phys Rev B 92:205134
17. Tao T, Ro J, Melngailis J, Xue Z, Kaesz HD (1990) Focused ion beam induced deposition of platinum. J Vac Sci Technol B 8:1826–1829
18. Moll PJW (2018) Focused ion beam microstructuring of quantum matter. Annu Rev Condens Matter Phys 9:147–162
19. Pfau H, Daou R, Friedemann S, Karbassi S, Ghannadzadeh S, Küchler R, Hamann S, Steppke A, Sun D, König M, Mackenzie AP, Kliemt K, Krellner C, Brando M (2017) Cascade of magnetic-field-induced Lifshitz transitions in the ferromagnetic Kondo lattice material $YbNi_4P_2$. Phys Rev Lett 119:126402
20. Moll PJW, Puzniak R, Balakirev F, Rogacki K, Karpinski J, Zhigadlo ND, Batlogg B (2010) High magnetic-field scales and critical currents in SmFeAs(O, F) crystals. Nat Mater 9(8):628–633
21. Moll PJW, Nair NL, Helm T, Potter AC, Kimchi I, Vishwanath A, Analytis JG (2016) Transport evidence for Fermi-arc-mediated chirality transfer in the Dirac semimetal Cd_3As_2. Nature 535(7611):266–270
22. Nandi N, Scaffidi T, Kushwaha P, Khim S, Barber ME, Sunko V, Mazzola F, King PDC, Rosner H, Moll PJW, König M, Moore JE, Hartnoll S, Mackenzie AP (2018) Unconventional magneto-transport in ultrapure $PdCoO_2$ and $PtCoO_2$. Nat Quantum Mater 3(66)
23. Moll PJW (2012) The role of anisotropy in iron-pnictides addressed by focused ion beam sample fabrication. PhD thesis, ETH Zurich

Chapter 3
On the Ballistic Motion of Electrons in $PdCoO_2$

Electron transport through metals and semiconductors can directly be investigated on a mesoscopic scale by fabricating a material into a microscopic device. In particular, if the size of the conductor is well bellow the electron scattering length, the electron transport is dominated by boundary interactions. This distinct, so-called ballistic transport regime is well studied in ultra-pure two-dimensional semiconducting electrons systems, as well as graphene and similar 2D materials. Using a FIB to microstructure as grown single crystals, ballistic devices are no longer limited to ultra-optimised thin film materials, but can be studied in compounds which cannot be manipulated by lithographic means. This approach opens exotic material classes for ballistic investigations that are only available as single crystals.

During the course of my doctoral research I have discovered that the ultra-pure metallic oxide $PdCoO_2$, for instance, exhibits strongly anisotropic in-plane ballistic behaviour governed by the underlying hexagonal Fermi surface. This restricts the electron motion into mainly three directions and significantly influences the physics of ballistic electron transport. In the following, I will demonstrate the consequences of directional ballistics by studying the in-plane anisotropy induced entirely by electron-boundary interactions in the ballistic regime. Moreover, the real space ballistic trajectories can directly be probed by bending the electrons onto cyclotron orbits in a magnetic field, which is a technique known as transverse electron focusing (TEF) and will subsequently be discussed in detail.

The example of $PdCoO_2$ demonstrates how new electronic behaviour in exotic materials arises form a non-trivial two-dimensional Fermi surface. Furthermore, the following chapter showcases how microstructured devices can be tailored to suit the specific requirements of the underlying physical question.

M. D. Bachmann, *Manipulating Anisotropic Transport and Superconductivity by Focused Ion Beam Microstructuring*, Springer Theses,
https://doi.org/10.1007/978-3-030-51362-7_3

3.1 The Delafossite Metal $PdCoO_2$

One of the greatest technical challenges in the field of quantum materials lies in the material refinement of single crystals and thin films. A great deal of effort and often years of hard work go into achieving materials with ultimate purity and as little disorder as possible. In elemental metals, careful annealing can result in high quality crystals with electron mean free paths of several millimetres [1]. The situation in complex materials is more challenging, because macroscopic faults (e.g. flux inclusions, impurity phases, voids) and microscopic defects (e.g. dislocations, vacancies, interstitial defects, stacking faults, grain boundaries, twin boundaries) can lead to a high degree of disorder, much of which cannot simply be annealed away.

It is therefore perhaps surprising to find the class of ternary compounds to include a remarkable example of extreme crystalline purity. The metallic oxides $PdCoO_2$ and $PtCoO_2$ belong to the large family of delafossite oxides, first discovered in the form of the semiconducting mineral $CuFeO_2$ and named in honour of the mineralogist Gabriel Delafosse. Like many other layered oxide structural series, the delafossite structure, with a stoichiometry of the general form ABO_2, can host a large variety of elements. Only considering transition metals, the lattice sites can be occupied by A = Cu, Pd, Ag, Pt and B = Sc, Cr, Fe, Co, Ni, Y and Rh respectively. Akin to most other layered oxides, which are typically insulators, semiconductors, or bad metals, most of the transition metal delafossites have been found to be semiconductors and many of them have been investigated for possible applications in the field of transparent conducting oxides (TCO) [2]. Moreover the delafossite family also hosts the most conductive oxides known. Amongst the metals the non-magnetic compounds $PdCoO_2$ [3, 4], $PtCoO_2$ [5], and $PdRhO_2$ [6], as well as the frustrated antiferromagnetic $PdCrO_2$ [7, 8] ($T_N = 37.5$ K) can be found. Their commonality is the two-dimensional sheets formed by the Pd or Pt layers, which gives rise to the remarkably high conductivity. An exception is the nickel-based delafossite $AgNiO_2$, which also shows metallic character [9], which is, however, associated with the NiO_2 layers. This fact is also reflected in the room temperature resistivity value, which is $\sim 10^3$ higher than that for the aforementioned compounds [10], on which we will focus from here on.

Many of the delafossites were first synthesized by Shannon, Prewitt, Rogers et al. who published a series of three papers in 1971 describing nine newly discovered compounds [11–13]. Although they reported the amazingly high in-plane conductivity values in $PdCoO_2$ and $PtCoO_2$ at room temperature, corresponding to resistivities of a few $\mu\Omega$cm, little subsequent work was published over the next 20 years. Only in the mid 1990s, when Tanaka et al. [3, 14] reported the first strongly anisotropic temperature dependence of the resistivity of $PdCoO_2$, was the interest in the ultra-pure metallic delafossites renewed. It took another 10 years until Takatsu et al. [4] grew $PdCoO_2$ with substantially increased residual resistance ratios of over 400, resulting in an ultra conductive metal. When compared to the best metals known (Fig. 3.1a) the delafossites are by far the least resistive oxides and they outperform the alkali metals. They are only second to the 3D elemental metals Ag, Cu and Au

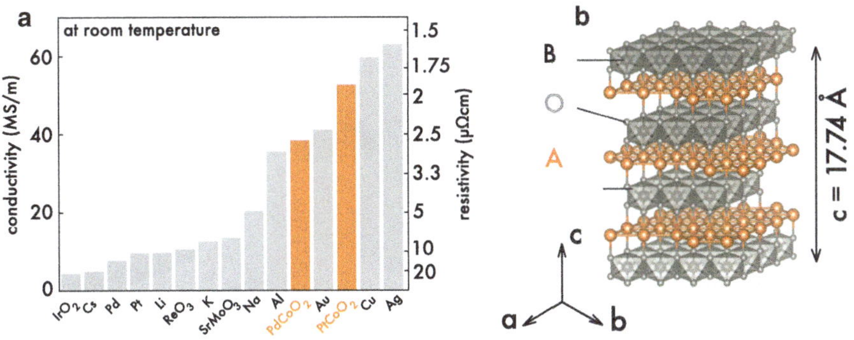

Fig. 3.1 **a** Comparison of the room temperature conductivities/resistivities of the highest conductivity metals known, including many of the Alkali metals, some low resistivity oxides and the well-known record holders copper and silver. Based on [15] and references therein. Values for $PdCoO_2$ and $PtCoO_2$ from [16]. **b** The delafossite crystal structure with the general formula ABO_2. In the case of $PdCoO_2$ palladium atoms form ultra mobile electron layers (orange) which are separated by insulating CoO_2 octahedral blocks

at room temperature, which however have a 3 times higher charge carrier density than $(Pd/Pt)CoO_2$. Since the revival of the delafossites by Takatsu et al., possibly as a result of its record length mean free path amongst all delafossites at low temperatures, the non-magnetic compound $PdCoO_2$ has been by far the most studied metallic delafossite and the subject of this chapter. On the following pages, I will give a brief overview of the crystal growth and physical properties of $PdCoO_2$.

$PdCoO_2$ crystallises in the trigonal space group $R\bar{3}m$ (166) (Fig. 3.1b) with lattice constants $a = b = 2.83$ Å and $c = 17.743$ Å [4]. The layered structure consists of single sheets of triangular coordinated palladium planes separated by insulating edge-shared CoO_2 octahedra. This forms a natural heterostructure of highly conductive two dimensional metal sheets separated by insulating layers. The unit cell contains 3 inequivalent Pd planes. The cobalt oxide delafossites are also reminiscent of the well known layered rock-salt $NaCoO_2$, which becomes superconducting when intercalated with water. $PdCoO_2$, however, is not superconducting (down to 15 mK [4]) and attempts at intercalating water have so far not been successful.

3.1.1 Crystal Growth and Characterisation

Astonishingly, the delafossite metals grow as ultra-pure single crystals without the necessity of any further purification. The single crystals are grown in an evacuated quartz ampoule with a mixture of the starting compounds following the metathetical reaction described in [4, 11]:

$$PdCl_2 + 2 \cdot CoO \rightarrow 2 \cdot PdCoO_2 + CoCl_2.$$

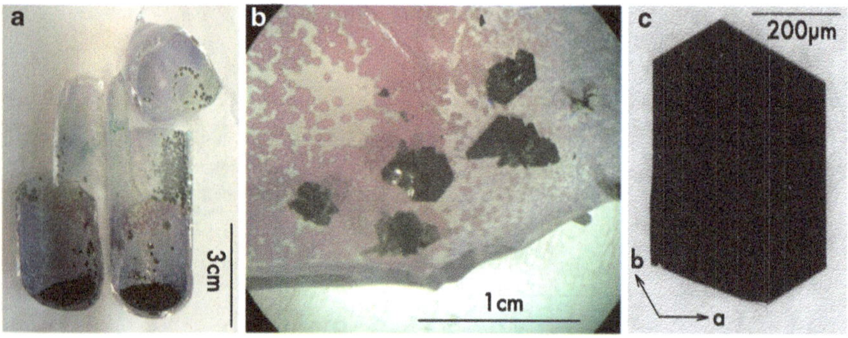

Fig. 3.2 Growth of delafossite single crystals. **a** Single crystal platelets have grown on the quartz wall slightly above the starting material at the bottom. **b** the $CoCl_2$ by-product (pink crystals) will be dissolved using distilled water and ethanol. No further purification techniques are needed. All delafossite crystals used in this work were grown by Seunghyun Khim. **c** A nearly perfect hexagonal single crystal platelet of $PdCoO_2$, about 500 μm long and 20 μm thick

The ampoule is heated to 1000 °C for 12 h and kept at 700–750 °C for further 5 days. Finally, to remove the excess $CoCl_2$, the resultant product is washed with distilled water and ethanol (Fig. 3.2). No further purification or refinement is necessary. For an oxide, this is particularly incredible given the extremely high quality of the final product (the mean free path at low temperatures is $l \sim 20\,\mu m$)! In comparison, years of careful material refinement have increased l to 1.5 μm in Sr_2RuO_4 [17] and only after encapsulating graphene in hexagonal boron nitride (hBN) have mean free paths of 20 μm and higher been achieved [18].

The result of the crystal synthesis is presented in Fig. 3.2. The high quality single crystals can be extracted from the growth ampoule by hand. Laue X-ray reflection shows sharp spots and determines the crystal axis to be perpendicular to the growth edges of the crystal as indicated. Using quantitative energy dispersive spectrometry the stoichiometric growth of the crystals is confirmed to within $\pm 1\%$ accuracy. (Fig. 3.3a).

Successful synthesis of highly conductive $PdCoO_2$ ultra-thin films was reported in early 2018 [19]. The simultaneous high values of electrical conductivity and optical transmittance make $PdCoO_2$ ultra-thin films promising transparent electrodes for technological applications. Based on pulsed laser deposition (PLD) this result opens the door to large scale industrial applications of delafossite oxides.

3.1.2 *Physical Properties*

Already in their initial work, Shannon et al. [11] noted that the unusual monovalent valence state of Pd leads to a linear coordination by anions (O^{2-}-Pd^{1+}-O^{2-}), which had not previously been observed for palladium. Subsequent magnetic susceptibility

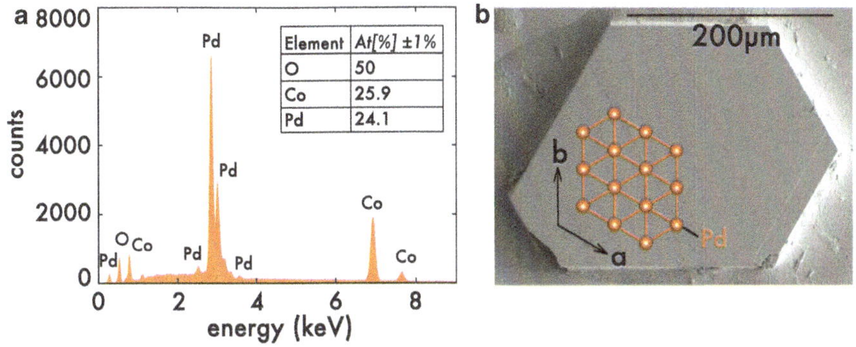

Fig. 3.3 **a** Quantitative energy dispersive spectrometry (EDS) analysis of the $PdCoO_2$ single-crystal platelet shown in panel b. The Oxford Instruments AZtec software platform was used to determine the elemental composition, which confirms the stoichiometric growth of the $PdCoO_2$. The errors on the elemental composition are ±3 At%. **b** The growth edges of the hexagonal platelets of $PdCoO_2$ are perpendicular to the crystal axis (c points out of the plane)

measurements indicated that the cobalt was in a trivalent, low-spin state Co^{3+}. This rare chemical environment forms the basis of the electronic structure of $PdCoO_2$.

Further insight was gained from early photoemission spectroscopy measurements (PES) [20–22] and ab initio band structure calculations [23], which established that the electronic density of states (DOS) at the Fermi level was dominated by Pd with very little contribution of Co. Additionally, specific heat measurements suggested that the Pd states are predominantly derived from s-d hybridised orbitals [24].

With the first angle resolved photoemission spectroscopy (ARPES) data, Noh et al. [25] could directly visualise the Fermi surface. They found a fragile, spin-split surface state and a nearly perfect hexagonal bulk FS, in good agreement with the bulk electronic structure predicted by density functional theory (DFT) calculations [26]. The authors concluded that the high electrical conductivity was due to a very dispersive s-d hybridised band, which generated a long carrier lifetime and large carrier velocity on the large 2D Fermi surface.

The ARPES measurements stimulated new theoretical studies to calculate the electronic structure. While previous DFT studies, such as those by Eyert et al. [27], agreed well with the experimental findings of a single hexagonal band of 2D nature, the band was mainly derived from Pd-4d orbitals. In 2010, however the DFT calculations by Ong et al. [28] for the first time also found a small Pd-5s contribution, in agreement with the experiments.

In 2012, Hicks et al. [29] presented de-Haas van Alphen (dHvA) magnetisation oscillation measurements from magnetic torque measurements. Three dHvA frequencies were found, corresponding to the extremal neck and belly orbits and their difference frequency. By performing an extensive rotation study and decomposing the Fermi surface into Fourier harmonics the 3D warping in the k_z direction was characterised. The corrugation of the FS suggests that the metallic out-of-plane transport stems from the overlap of extended Pd-5s orbitals. Also the dHvA masses

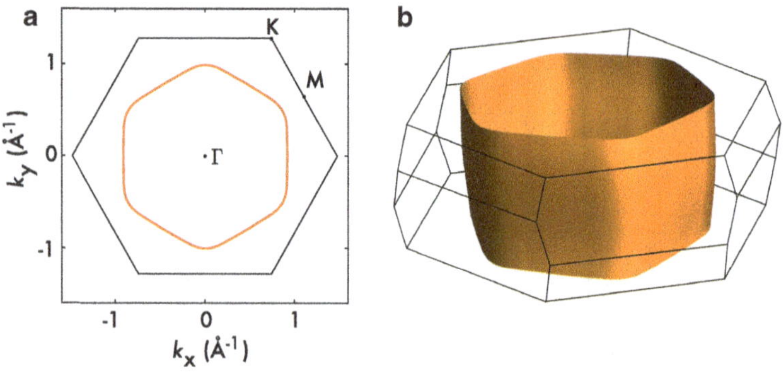

Fig. 3.4 The Fermi surface of $PdCoO_2$ from fitting a harmonic expansion to ARPES and dHvA measurements. Parameters found in [25, 29]. **a** The in-plane Fermi surface consist of a single, half-filled band with nearly hexagonal cross-section. **b** The nearly cylindrical FS has a small degree of warping, evident by the metallic c-axis behaviour

of $\approx$1.5 m_e were too low to be originating from localised d-orbital bands, but could be well explained in the picture of s-orbital contribution to the conduction band. In a DFT calculation Hicks et al. found that a $U = 5\,$eV cobalt on-site repulsion must be added to agree with the experimentally determined electron masses and FS warping. However, as the conduction electron masses mainly originate from the Pd-Pd bonds the involvement of cobalt layer came as a surprise. The authors suggest that this reflects the extended Pd-5s orbitals which sense the charge distribution in the CoO_2 layer. This charge interplay between the CoO_2 layer and the interstitial planes is reminiscent of the research in the related layered rock-salt compound Na_xCoO_2, where Coulomb interactions imprint a Na charge landscape onto the CoO_2 layer [30].

The inspection of the large flat sections on the Fermi surface in Fig. 3.4 naturally leads to the question of density wave formation—or rather absence thereof. No signatures of density waves have been reported although the extended parallel sections of the nearly perfect hexagonal FS at half-filling suggest ideal nesting conditions. The associated wave vector, however, is incommensurate, and the nearly hexagonal FS is stable with no signs of gapping down to 23 mK.

Hicks et al. further performed high resolution resistivity measurements on bulk crystals. They highlight the extremely low in-plane residual resistivity below 20 K, $\rho_{ab} \approx 8.6\,$nΩcm and high residual resistivity ratio, RRR $\approx$400 of $PdCoO_2$. This corresponds to a scattering event one in every $\sim 10^5$ lattice spacings at low temperatures, which suggests that either the Pd layers are almost completely free of disorder or that an unusual mechanism which reduces electron scattering by impurities might be at play. A recent theoretical study on the closely related compound $PtCoO_2$ [31] argues that strong orbital-momentum locking around the Fermi surface suppresses the in-plane electron scattering rate τ by roughly a factor 3. While this may explain the low value of the room temperature resistivity, the underlying reason for exceptionally clean crystal growth, which leads to an unusually low residual resistivity, remains an open question.

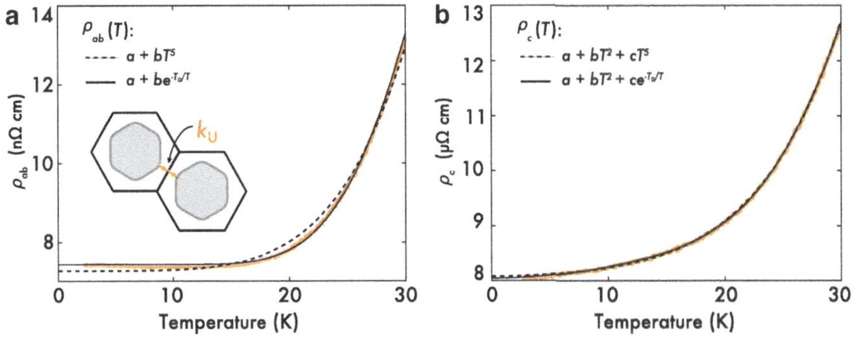

Fig. 3.5 Low temperature resistivities in $PdCoO_2$. While in **b** the out-of-plane transport follows the standard Bloch–Grüneisen T^5 behaviour, the **a** in-plane resistivity does not. Instead it can be fitted to an activated form (solid line) with $T_0 = 157\,K$. Reproduced from [29]

The low temperature measurements also reveal an unusual temperature dependence of the in-plane resistivity ρ_{ab} (Fig. 3.5a). It cannot be fitted to the standard $\propto T^5$ Bloch–Grüneisen dependence describing electron-phonon scattering. Instead an activated behaviour of the form $a + be^{-T/T_0}$ with $T_0 = 157\,K$ (solid line in Fig. 3.5a) describes the data well. The out-of-plane transport can be fitted with the standard $a + bT^2 + cT^5$ form, which takes into account the standard electron-electron scattering Fermi liquid T^2 term, as well as the T^5 term due to electron-phonon scattering.

One explanation for the activated behaviour of the in-plane resistivity may be phonon drag. If this is the case, the phonon population is pushed out of equilibrium by the electron flow and is unable to re-equilibrate. As a result normal electron-phonon scattering transfers momentum between electrons and phonons but does not relax the momentum of the combined system and therefore will not contribute to the resistance. The lowest order electron-phonon processes that will relax momentum and affect the resistance will be electron-phonon Umklapp processes in which a phonon scatters an electron to the next Brillouin zone. In a material with a single, closed Fermi surface sheet, there is an associated energy gap $k_B T_U = \hbar c_s k_U$, where k_U is the minimum wave number for the Umklapp process and c_s is the sound velocity. Careful thermopower measurements by Daou et al. [32] in 2015 support this hypothesis by reporting a dip in the Seebeck coefficient around 50 K, where phonon drag is the strongest.

High precision measurements on FIB microstructured samples across the entire temperature range confirm the strongly two-dimensional electronic character, yielding a resistive anisotropy of $\rho_c/\rho_{ab} > 500$ above 50 K and over 2500 below 20 K. In simple metals, where the resistivity is governed by electron-acoustic phonon interactions, the Bloch–Grüneisen law predicts a T-linear temperature scaling well above the Debye temperature. In $PdCoO_2$, the values of Θ_D obtained from resistivity are slightly above room temperature, $\Theta_D = 340\,K$ for ρ_{ab} and $\Theta_D = 310\,K$ for ρ_c [4]. This suggests that the observed super-linear resistivity behaviour between $\sim$100 K and 300 K is because the system is in a crossover regime.

3.1.3 The Mean Free Path

The elastic mean free path l of the electrons can be extracted from the in-plane resistivity ρ_{ab} using the standard 2D expression for circular Fermi surfaces

$$\rho_{ab}^{-1} = \frac{e^2}{hd} k_F l, \tag{3.1}$$

where $d = 17.73$ Å/3 is the interlayer separation and k_F is the average Fermi momentum defined by $k_F^2 = 2\pi n$. The derivation of this expression is given at the end of this Sect. 3.1.4.

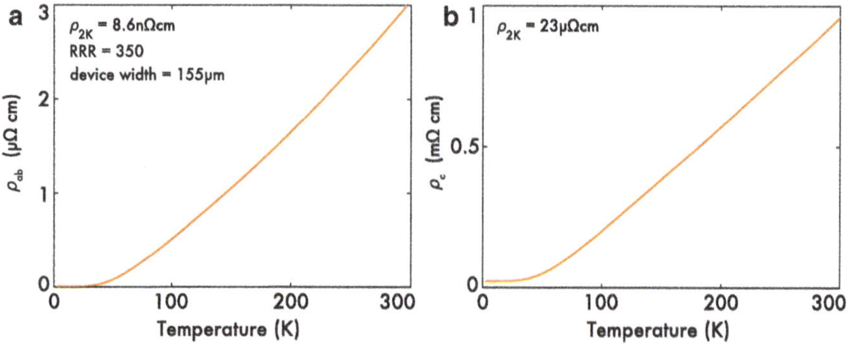

Fig. 3.6 In-plane and out-of-plane resistivity of $PdCoO_2$. **a** The in-plane resistivity ρ_{ab} from 2 to 300 K was measured on a FIB microstructured crystal platelet cut into a 155 µm wide channel. Data kindly provided by N. Nandi. **b** The out-of-plane resistivity ρ_c over the same temperature range also shows a metallic behaviour. The c-axis measurement was performed on a microstructured lamella, which provides excellent control over the current path, such that in-plane contributions are minimised

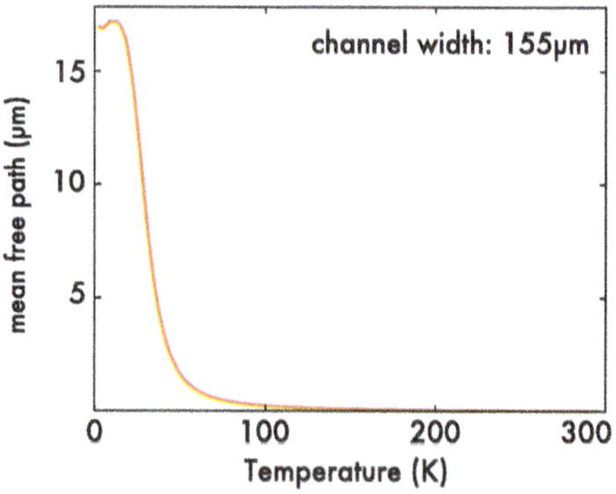

Fig. 3.7 Extraction of the electron mean free path l in $PdCoO_2$ using the standard 2D expression $\rho_{ab}{}^{-1} = \frac{e^2}{hd} k_F l$. The resistivity (c.f. Fig. 3.6a) was carefully measured in a 155 µm wide microstructured channel, which is assumed to be in the diffusive limit

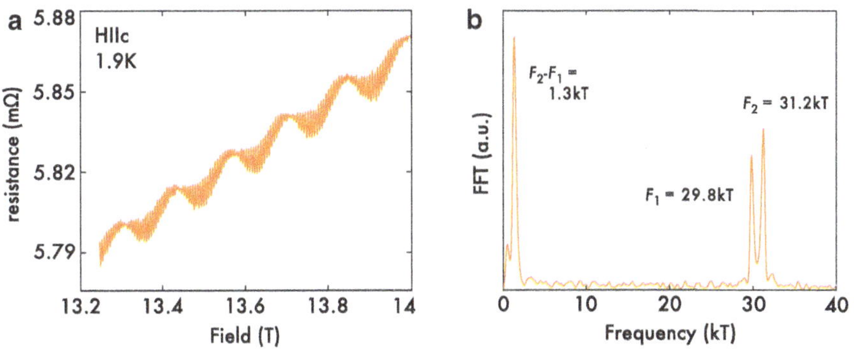

Fig. 3.8 Shubnikov de Haas oscillations measured at H||c and 1.9 K in a FIB fabricated device. **a** Bare SdH oscillations after a smooth background was subtracted. The high frequency oscillations are well visible and correspond to the two bulk frequencies around 30 kT. **b** The fast Fourier transform (FFT) of the signal in A shows the two main frequencies F_1 and F_2 and their difference frequency, in agreement with dHvA measurements [29]

Using measurements from a $w = 155\,\mu$m wide transport bar (Fig. 3.6a, from [16]), which is assumed to be in the diffusive transport regime we can then extract the mean free path l over the entire temperature range, see Fig. 3.7. At room temperatures, the mean free path is approximately 50 nm and slowly increases to about 300 nm at 100 K. Around 50 K it dramatically rises and reaches a value of 17 μm. The temperature scale of the rapid increase of the mean free path coincides with the dip in the Seebeck coefficient, associated with strong phonon drag, which suggests an interdependence of these phenomena.

Quantum oscillations, which are the result of the Landau quantisation of charge carrier orbits in a magnetic field, are considered one of the most stringent tests of crystalline quality. The amplitude is moderated by $e^{-\pi r_c/l_D}$, an exponential function depending on the so-called Dingle mean free path l_D and the cyclotron radius $r_c = \frac{\hbar k_F}{eB}$ which exponentially damps the detectable oscillations. It therefore comes as no surprise that quantum oscillations are readily observable in the delafossites up to 2 K (Fig. 3.8). Surprisingly, in contrast to the 20 μm mean free path which is extracted from resistivity measurements, the Dingle mean free path which is extracted from dHvA (600 nm) [29] or SdH measurements (660 nm) is about a factor of 30 smaller than the resistive mean free path. An important difference between these length scales is that the quantum oscillation based analysis is sensitive to phase coherent processes, while the resistive mean free path is not. Another key difference is that a dHvA or SdH experiment samples the entire FS by forcing the electron to revolve around it. The Drude analysis of the mean free path is performed in the absence of magnetic fields and does not require the charge carriers to experience all parts of FS equally. A scenario where some parts of the FS (for example the corners of the hexagons) scatter electrons more strongly than others could lead to the observed phenomenology.

Finally I close this section by comparing the physical properties of $PdCoO_2$ to an ultra-mobile 2DEG hosted by a GaAs/AlGaAs heterostructure, on which a study

Table 3.1 Comparison of the electronic properties of $PdCoO_2$ (data from [15] and references therein) and the 2DEG in a GaAs/AlGaAs heterostructure (data from [33])

		$PdCoO_2$	GaAs (100)	Units
Effective mass	m^*	1.49	0.067	$m_e = 9.1 \cdot 10^{-31}$ kg
Electron sheet density	n_s	$1.45 \cdot 10^{15}$	$9.1 \cdot 10^{11}$	cm^{-2}
Fermi wave vector	$k_F = \sqrt{2\pi n_s}$ †	0.96	0.024	$Å^{-1}$
Fermi wavelength	$\lambda_F = 2\pi/k_F$	6.5	260	Å
Fermi energy	$E_F = (\hbar k_F)^2/2m^*$	27270	377	K
Sheet resistivity	ρ_{2D}	0.126	2.7	Ω/sq
Scattering time	$\tau = m^* \mu_e / e$	29	95	ps
Fermi velocity	$v_F = \hbar k_F / m^*$	7.5	4.1	10^7 cm/s
Mean free path	$l = v_F \tau$	21	40	μm

†In a system with spin degeneracy $g_s = 2$ and valley degeneracy $g_v = 1$.

of electron hydrodynamics was recently reported [33]. Despite having comparable mean free paths l and similar scattering times, these materials are very different (c.f. Table 3.1). $PdCoO_2$ is a monovalent metal with a high electron density, has a large Fermi surface at half filling (large k_F) and very fast electrons. The 2DEG in GaAs/AlGaAs is a low density semiconductor with a much smaller FS and much lighter electrons. The low charge carrier density implies a large Fermi wavelength, comparable to the smallest dimensions of the structure. One of the key advantages of the semiconductor is the large variability of the electron density which may be tuned by means of an electric field (i.e. the gate voltage).

Although it is a blessing for reaching regimes of yet unexplored ultra-pure high-density electron transport, the ultra-low, highly anisotropic resistivity of $PdCoO_2$ poses a challenge for any experimentalist. Fortunately, the technology now exists to perform precise and controlled measurements on it. By using FIB technology, well-defined channels which are perfectly oriented along a specific crystalline direction can be fabricated. Moreover, the capability of the FIB to precisely control the shape and size of a 3D crystals has led to the investigation of viscous electronic flow [34] in $PdCoO_2$. In the work presented in this chapter, I used the FIB to reduce and confine the sample dimension to well below the mean free path of the electrons to investigate the unusual ballistic effects. The rest of the chapter is divided into 3 parts: I will first give an overview of ballistic transport phenomena observed in clean elemental metals and high-mobility two-dimensional systems. This sets the framework for the remaining sections in which I present my ballistic transport experiments in FIB defined $PdCoO_2$ crystals.

3.1.4 Derivation of the Two-Dimensional In-Plane Conductivity

Here I derive expression (3.1), which relates the in-plane conductivity $\sigma_{ab} = \rho_{ab}^{-1}$ (in zero-magnetic field) to the mean free path l.

$$\rho_{ab}^{-1} = \frac{e^2}{hd} k_{\mathrm{F}} l,$$

This form can be derived from the expression for the conductivity given in [35] for a single band metal

$$\sigma = \frac{e^2}{4\pi^3 \hbar} \iint_{S_{\mathrm{F}}} \frac{\tau(\vec{k}) v_t^2(\vec{k})}{|\vec{v}(\vec{k})|} dS. \tag{3.2}$$

In general, the electron relaxation time $\tau(\vec{k})$, the electron velocity along the transport direction $v_t(\vec{k})$ and the electron velocity $\vec{v}(\vec{k}) = \frac{1}{\hbar} \nabla_{\vec{k}} E(\vec{k})$ are functions of the wave vector $\vec{k}$, which must be integrated over the Fermi surface S_{F}. Using a constant mean free path approximation, the carrier relaxation time becomes $\tau(\vec{k}) = l/|\vec{v}(\vec{k})|$ and the expression in (3.2) can be written as

$$\sigma = \frac{e^2 l}{4\pi^3 \hbar} \iint_{S_{\mathrm{F}}} \frac{v_t^2(\vec{k})}{\vec{v}^2(\vec{k})} dS = \frac{e^2 l}{4\pi^3 \hbar k_{\mathrm{F}}^2} \iint_{S_{\mathrm{F}}} k_x^2 \, dS,$$

where in the second step the parabolic band approximation ($v(\vec{k}) = \frac{\hbar}{m*} \vec{k}$) was used to simplify the integrand ($\frac{v_t^2(\vec{k})}{\vec{v}^2(\vec{k})} = \frac{k_x^2}{k_{\mathrm{F}}^2}$). Without loss of generality the direction of transport can be chosen to be along the x-axis. To perform the surface integral I approximate the Fermi surface of the two-dimensional system with a cylinder of height $\frac{2\pi}{d}$ ($d = \frac{c}{3}$ is the palladium interlayer separation) and radius k_{F}. A cylindrical coordinate transformation yields $k_x = k_{\mathrm{F}} \cos(\theta)$, $k_y = k_{\mathrm{F}} \sin(\theta)$, $k_z = z$, and $dS = k_{\mathrm{F}} \, d\theta \, dz$, with $0 \leqslant \theta \leqslant 2\pi$ and $0 \leqslant z \leqslant 2\pi/d$, such that the surface integral can be easily evaluated

$$\iint_{S_{\mathrm{F}}} k_x^2 \, dS = \int_0^{2\pi} k_{\mathrm{F}}^2 \cos^2(\theta) \cdot k_{\mathrm{F}} \, d\theta \int_0^{2\pi/d} dz = k_{\mathrm{F}}^3 \pi \frac{2\pi}{d}.$$

Gathering all the terms we find the expression fo the conductivity

$$\sigma = \frac{e^2 l}{4\pi^3 \hbar k_{\mathrm{F}}^2} k_{\mathrm{F}}^3 \pi \frac{2\pi}{d} = \frac{e^2}{hd} k_{\mathrm{F}} l.$$

3.2 Electron Conduction in the Ballistic Transport Regime

3.2.1 *Length Scales and Their Significance*

In general, there are multiple intrinsic length scales which characterise charge transport through a conductor depending on their size relative to the sample size L. The following table (Table 3.2) by Hashimoto et al. [36] classifies electronic transport into different regimes, depending on the size of the Fermi wavelength λ_F, the phase coherence length l_φ and the bulk mean free path l.

In $PdCoO_2$, λ_F is 0.65 nm, i.e. approximately 2 lattice constants. This is a consequence of the fact that every Pd atom contributes one conduction electron, leading to a half-filled band crossing the Fermi level. Confining the two-dimensional in-plane transport to only 1D is hence not feasible and so $PdCoO_2$ falls into the dimensionally unrestricted category. As emphasised above, the mean free path l reaches up to $\sim$20 μm at low temperatures. With the ability of the FIB to tune the sample dimensions from hundreds of microns down to hundreds of nanometers, both diffusive and ballistic effects can be studied.

Table 3.2 Classification of electronic transport regimes according to their characteristic length scales. λ_F is the Fermi wavelength, l is the bulk mean free path and l_φ is the length over which the phase of the electron remains coherent. Table from [36]

x	$L > x$	$L < x$
λ_F	No dimensional restriction	Dimensional restriction
l	Diffusive	Ballistic
l_φ	Macroscopic (semi-classical)	Mesoscopic (quantum)

3.2.2 *From Diffusive to Ballistic Transport*

Typically, electrons in metals undergo frequent collisions with impurities, phonons and other electrons which limit the bulk mean free path l to well below the dimensions of the sample L ($l \ll L$). This limit is called the diffusive transport regime and is well described by Ohm's law $\vec{J}(\vec{r}) = \sigma \vec{E}(\vec{r})$. It states that the local current density $\vec{J}(\vec{r})$ can be related to the generated electric field $\vec{E}(\vec{r})$ at any given location $\vec{r}$, through the material-specific conductivity tensor σ. In particular, σ is an intrinsic quantity, which describes the ability of a material to conduct current independent of its geometry and dimensions. The number of independent elements of the second rank conductivity tensor (i.e. a 3×3 matrix) can be reduced according to the symmetries of a system. The individual components of σ are determined experimentally via resistance measurements, $R = V/I$, which relate the measured voltage drop V to the

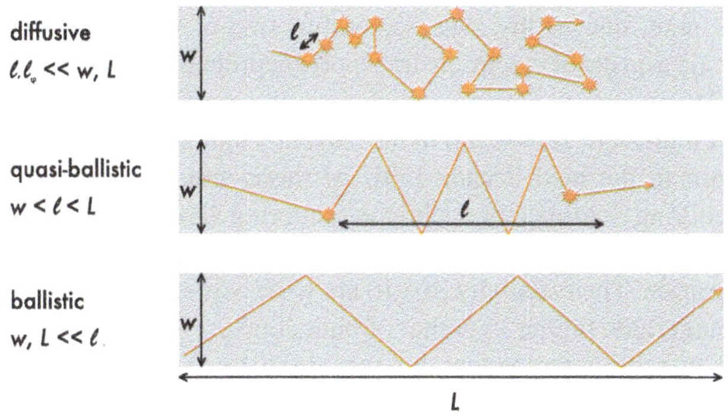

Fig. 3.9 Classification of transport regimes in a channel with perfectly specular reflecting boundaries. In the diffusive transport regime the sample dimensions are much greater than the electronic mean free path. If the scattering events are elastic, then the phase-coherent mean free path l_φ can be longer than the mean free path l. The opposite limit, where the sample dimensions are much smaller compared to the mean free path, is called the ballistic regime. The case where the mean free path of the electrons is larger than the channel width, but shorter than the channel length is termed quasi-ballistic. Figure based on [38]

applied current I. For arbitrary sample geometries, the determination of the relation between R and the resistivity ρ typically requires the use of finite element analysis tools, however for very simple geometries the problem can be greatly simplified. For instance, in the case of a long transport bar, the solution is given by the well known expression $\sigma^{-1} = \rho = R\,\frac{t \cdot w}{l}$ (in absence of a magnetic field), where t, w and l are the sample thickness, width and length respectively. This entire analysis breaks down when the mean free path becomes comparable to, or larger than, the system dimension, in which case the conductivity tensor σ is no longer a well-defined quantity.

The ballistic transport regime, in which the electronic mean free path is much longer than the sample dimensions ($l \gg L$), can be reached when a material is made either ultra pure or very small. In the ballistic limit, boundary collisions are the main source of resistance, while electron scattering from impurities, phonons and electrons can mostly be neglected. The ballistic regime is subdivided into a *quasi-ballistic regime*, where only one dimension of the system is smaller than the l and in to the fully *ballistic regime*, where the sample dimensions are the only relevant length scales (Fig. 3.9). According to this classification most of the experiments I will discuss are performed in the quasi-ballistic regime, i.e. a small but finite amount of bulk scattering is present. In the ballistic limit, the concept of "conductivity", and equally "resistivity", lose their original meaning, as they no longer reflect an intrinsic material property but depend heavily on the geometry and dimensions of the sample. So instead of solving the electrostatic problem, the analogy of electrons and their trajectories as semiclassical "billiard balls" which move on the Fermi surface with velocity v_F and scatter from the boundaries is more appropriate [37].

On a side note, due to the ever decreasing size of integrated microelectronic circuits, ballistic aspects of electronic transport are probably more relevant today than ever before. The copper nanowires which form the metallic interconnect structures in integrated circuits are sub-30 nm in the current technology node and are expected to reach 10 nm in the next decade [39]. At these dimensions, surface and grain boundary scattering dominate over phonon scattering at room temperature, resulting in enhanced resistivity values and accelerated deterioration of the interconnects due to electromigration. The resistivity of a 10 nm wide copper interconnect line is about an order of magnitude higher than that of bulk copper [40], which presents a major challenge for metallic interconnect circuitry. In 2013 the International Technology Roadmap for Semiconductors (ITRS) declared the introduction of new materials that meet wire conductivity requirements to be their biggest near term challenge for interconnects. This became even more explicit in their 2015 report, which specifies the need to replace copper as interconnect material to limit the resistance increase at reduced scale in order to minimise both power consumption and signal delay. Ballistic effects are also of concern for related fields of technology which operate on small length scales, such as transparent flexible conductors [41], spintronics devices [42] or magnetic sensors [43].

In the remaining parts of this section I will review ballistic transport phenomena observed in (i) clean metals and (ii) semiconductor heterostructures. Due to the small Fermi wave length λ_F, the size effects in metals are classical, whereas in semiconductors both classical and quantum size effects occur. This subdivision is also temporal, as most of the research on metals was done between 1900 and 1990 and was nearly completely superseded by the arrival of clean semiconductor two-dimensional electron gas systems (2DEGs) and, most recently, by high-quality graphene, in which mean free paths of many microns have been achieved [18]. As a complete overview of the countless ballistic effects observed in semiconducting structures goes well beyond the scope of this thesis, I refer the interested reader to an excellent review by Beenakker and van Houten [37] as well as two excellent books on mesoscopic transport by Datta [44] and Ihn [45]. Instead I intend to give an overview of the relevant ballistic phenomena and formalisms to set my work on $PdCoO_2$ in context.

3.2.3 *Classical Size Effects in Metals*

In 1901 Thompson [46] correctly proposed that the additional resistance reported in thin films of silver compared to bulk samples [47] had its origin in collisions of conduction electrons with the boundary. The first and perhaps best known theoretical work to quantify this phenomenon was carried out by Fuchs [48] and Sondheimer [49] for plate geometries (thin films) and by Dingle [50] for cylindrical wires. Their calculations are based on an extension of the Boltzmann equation, in which a boundary condition is imposed on the electron distribution. By assuming a spherical Fermi surface and introducing a phenomenological specularity parameter p, which quantifies the probability for electrons being elastically reflected by the surface, they predict a resistivity of the form

$$\frac{\rho}{\rho_{bulk}} = 1 + \frac{3}{8}\frac{l}{L}(1 - p) \tag{3.3}$$

in the limit of $l \gg L$. Although widely applied due to its simplicity, this model neglects the variations in surface specularity due to surface roughness. Soffer [51] takes this into account by introducing an angle dependence of the reflection probability in his calculations. Since then many more theories have been developed, notably by Chambers [52], who uses a model based on kinetic equations and MacDonald [53], who calculates the resistivity of thin films in transverse magnetic fields. For widely used polycrystalline films, Mayades & Schatzkes investigated the effects of specular [54] and arbitrary [55] scattering from grain boundaries.

Providing reliable experimental values to compare to theoretical predictions was problematic at first, as producing samples over a wide range of thicknesses without influencing the sample quality was experimentally challenging. For instance, while thick films were produced by rolling bulk metals into foils, thin films were fabricated by evaporation. In the later case, interactions with a substrate could lead to cluster formation ('islandization') and result in a much rougher surface [56], such that a direct comparison between results from thick foils and thin films was problematic.

The quality of the experimental work improved significantly with the availability of low temperature measurements. At cryogenic temperatures, the bulk mean free path l is greatly enhanced compared to its room temperature value, which meant that boundary effects could be studied in thicker, higher quality samples. This becomes apparent upon consideration of Matthiessen's rule, which states that resistivities from independent scattering events can be treated as a parallel resistor network [57]. The bulk mean free path is then given by

$$\frac{1}{l} = \frac{1}{l_{e-e}} + \frac{1}{l_{e-ph}} + \frac{1}{l_{e-imp}},$$

where l_{e-e} is the electron-electron, l_{e-ph} is the electron-phonon and l_{e-imp} is the electron-impurity scattering length respectively. At room temperature, the resistivity is typically dominated by electron-phonon scattering, which limits l to below 100 nm in most materials [58]. Upon cooling, electron-electron collisions fall off as $\propto T^2$ and electron-phonon scattering declines $\propto T^5$ at low temperatures (Bloch–Grüneisen), such that eventually l is dominated by the impurity concentration, which, for very pure metals, can yield a mean free path of several millimetres [36].

No rule is without exception and in fact a great amount of effort, both theoretically and experimentally [59], has gone into studying deviations from Matthiessen's rule in clean metals due to size dependent effects. One possible scenario worth mentioning can be qualitatively understood by considering the impact of small angle electron-phonon scattering. At low temperatures in bulk metals, these are the dominant form of electron-phonon interaction and hence many collisions are necessary to remove momentum from an electron. On the other hand, in thin samples, a few phonon collisions may be enough to guide an electron originally moving parallel to the sample boundary into the surface where diffusive scattering may occur. Hence phonons are particularly effective in reducing the mean free path of electrons which contributed significantly to the conductivity and will hereby create a stronger temperature dependence in thin samples than they do in thicker samples.

In conclusion, many theories have been formulated to explain the effects of size depending boundary scattering effects in clean metals. Although they qualitatively predict the observed phenomena, none of them is in satisfactory quantitative agreement with a wide range of experiments. This comes as no surprise, as we have discussed that microscopic details of the boundaries are important in the ballistic regime. This includes the details in the preparation of the boundaries (surface roughness, amount and angle dependence of specularity) as well as details of underlying anisotropic phonon spectrum and electron distribution. Many of these factors are better understood in semiconductor 2DEGs with circular Fermi surfaces and well-defined boundary conditions.

Finally, as mentioned above, we have only considered semiclassical ballistic effects in pure metals. If the phase coherence length l_φ becomes long enough, the electronic wave function will become sensitive to interference phenomena, such as weak localisation [60], which can be observed in disordered thin films [61], and universal conductance fluctuations [62]. Ultimately when the sample size is restricted to the order of only a few nanometers, for instance in metallic nanoparticles or very thin films (<50 nm), electron confinement will result in the discretisation of energy levels, leading to observable quantum size effects [63]. These effects will either be discussed in the following section or go beyond the scope of this thesis.

3.2.4 Size Effects in Semiconductor Heterostructures

Artificially created two-dimensional electron gases (2DEGs) have become the model system of choice for studying mesoscopic physics due to their high crystalline purity and their compatibility with lithographical fabrication techniques. Semiconductor 2DEGs are commonly made from modulation doped GaAs/AlGaAs heterostructures and consist of a single, two-dimensional layer of highly mobile electrons on a circular Fermi surface. Electrostatic gates above the 2DEG can locally tune the electron density and partially or completely deplete or fill the electronic system. In this way the boundaries of a 2DEG are typically electrostatically defined. Due to their two-dimensionality, 2DEGs have some commonality with $PdCoO_2$, which can be thought of as a natural heterostructure of many thousands of two-dimensional palladium layers. However many properties of these systems are very distinct, most notably the small Fermi temperature and large Fermi wavelength λ_F in 2DEGs. A side-by-side comparison of the properties of $PdCoO_2$ and a GaAs/AlGaAs 2DEG can be found in Table 3.1. In the following I will review some of the mesoscopic phenomena occurring in semiconducting 2DEG systems and assess if we expect to see similar effects in $PdCoO_2$.

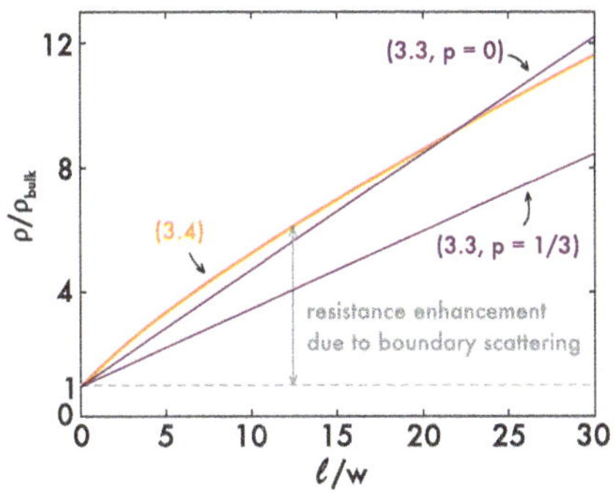

Fig. 3.10 Resistivity enhancement due to boundary scattering in two-dimensional systems with circular Fermi surfaces and diffusive boundaries based on Eq. (3.4). Additionally, the expression for thin 3D films (Eq. 3.3) with surface specularity parameters $p = 0$ and $p = 1/3$ are displayed in purple

Much like in three dimensional metals, the channel resistance of a 2DEG increases with decreasing channel width w due to an increase in boundary collisions. Assuming diffusive boundaries and a circular Fermi surface the Boltzmann equation can be used to calculate the dependence of the resistivity enhancement ρ/ρ_{bulk} as a function of the ratio of the momentum relaxing mean free path l and the channel width. The resulting expression derived in [37] is

$$\frac{\rho}{\rho_{bulk}} = \left[1 - \frac{4l}{\pi w} \int_0^1 \xi \sqrt{1-\xi^2}\,(1 - e^{-w/(l\xi)})d\xi\,\right]^{-1} \tag{3.4}$$

where the integral in Eq. (3.4) can be easily evaluated numerically. The corresponding functional dependence for the resistivity enhancement ρ/ρ_{bulk} is plotted in Fig. 3.10 alongside the form for isotropic 3D thin films with diffusive ($p = 0$) and partially specular boundaries ($p = 1/3$) given in Eq. (3.3).

As we will see in later sections, the hexagonal Fermi surface of $PdCoO_2$ will give rise to deviations from an isotropic system described by (3.4). Depending on the direction in which ballistic transport is measured, the flat Fermi surface facets can cause an increase or decrease of ρ/ρ_{bulk} in $PdCoO_2$ (c.f. Sect. 3.3.3). Moreover, the effective surface specularity measured in FIB defined $PdCoO_2$ channels benefits from a geometrical enhancement, again due to the nearly perfect hexagonal Fermi surface shape (c.f. Sect. 3.4.7), which leads to a suppression of the value of ρ/ρ_{bulk} as indicated in Fig. 3.10. These subtleties should be kept in mind when discussing classical size effects in non-circular Fermi surface systems.

A further way of probing size effects in metals is by applying a magnetic field and thereby introducing a further length scale, the cyclotron radius r_c, which can be conveniently tuned by adjusting the magnetic field strength $r_c \propto 1/B$. A typical experiment for measuring magneto-size effects consists of a constricted channel of width w in an out-of-plane magnetic field, such that the electrons undergo cyclotron motion in the plane of the 2DEG. Figure 3.11a displays the calculated resistivity increase ρ/ρ_{bulk}

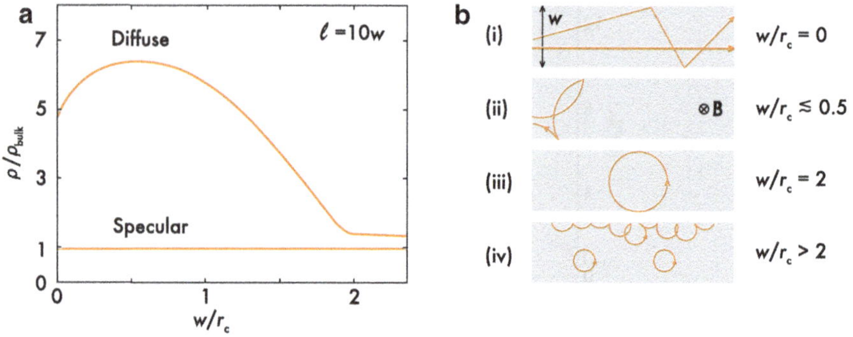

Fig. 3.11 Magneto-transport through a narrow channel of an isotropic 2D system. **a** The resistivity enhancement ρ/ρ_{bulk} for diffusive and specular boundaries is shown as a function of magnetic field $\propto 1/r_c$. Figure adapted from [38]. **b** Evolution of the electron orbit with increasing magnetic field. The resistivity response drastically changes above $w/r_c = 2$, when the cyclotron diameter is smaller than the channel width

of a constricted, isotropic 2D system for diffusive and specular boundaries. The calculations are based on solving the stationary Boltzmann equation in a magnetic field. In zero field, specular scattering does not affect the resistivity because the projection of the motion of the electron into the channel is not affected by the boundaries. It can be shown [38] more generally that the longitudinal resistivity is independent of a magnetic field for a system with specular boundaries. In the case of diffusive boundaries, the zero-field value (Fig. 3.11b(i)) is given by (3.4). Upon applying a magnetic field an initial increase of the resistivity is observed due to the deflection of electrons towards the boundary with a velocity nearly parallel the channel axis (Fig. 3.11b(ii)). The resistance reaches a maximum when the electrons are guided into the boundaries most efficiently, which is roughly at $w \approx 0.5l$ in a system with circular Fermi surface (Fig. 3.11b(iii)). Once the cyclotron diameter is smaller than the width of the channel, diffusive boundary scattering cannot reverse the direction of motion along the channel as it could for smaller magnetic fields (Fig. 3.11b(iv)). Beyond this point the resistivity becomes essentially field independent.

A different mechanism which can increase the resistivity value through back scattering in a ballistic system is weak localisation. This effect can occur in large diffusive metals with phase coherent lengths greater than the elastic mean free path $l_\varphi > l_e$, but smaller than the system size $l_\varphi < L$. In brief, weak localisation can be traced back to the constructive quantum interference of time-reversed paths of electrons. These lead to coherent back scattering in a channel and give rise to an anomalous increase of the resistivity with decreasing temperature. Moreover weak localisation is suppressed by a magnetic field, as it breaks time-reversal invariance. The magnitude of the weak localisation correction $\Delta\sigma_{wl}$ to the Drude σ conductivity for a two-dimensional system in zero-field is on the order of [45]

$$\frac{\Delta\sigma_{wl}}{\sigma} \approx -\frac{1}{k_F l_e} \ln\left(\frac{l_\varphi}{l_e}\right). \tag{3.5}$$

For a half-filled metal with a large k_F, such as the delafossites, the ratio of $\frac{l_\varphi}{l_e}$ would need to be enormous to see any effect at all. Hence we do not expect to see corrections arising from weak localisation in $PdCoO_2$. I mention this in particular in light of the tiny but anomalous upturn in the resistivity sometimes observed in $PdCoO_2$ at low temperatures, which therefore can not be explained by weak localisation.

Other interference phenomena occurring in ballistic systems with long phase coherence lengths l_φ are aperiodic magneto-conductance fluctuations and, in ring geometries, periodic Aharnonov–Bohm (AB) oscillations [64, 65]. These effects are likewise based on coherent backscattering of electrons which constructively interfere with themselves. In the case of magneto-conductance fluctuations, the oscillations depend on the local impurity potential of a device and causes a sample to sample variation of the periodicity of the oscillations. In contrast, the periodicity of AB oscillations is well-defined by an area S, which is threaded by a flux Φ leading to periodic oscillation with frequency $\Delta B = \frac{h}{e}\frac{1}{S}$. In practice, the first harmonic $\Delta B = \frac{h}{2e}\frac{1}{S}$ frequency is more often observed, as so-called Altshuler–Aronov–Spivak (AAS) oscillations. As these contain contributions from counter-propagating time-reversed paths which lead to constructive interference, AAS oscillations are more robust against self-averaging as the fundamental AB periodicity. These phenomena are in principle observable when the phase coherence length l_φ is larger or on the order of the system size and hence we might expect to see quantum interference effects in $PdCoO_2$ under the right conditions.

In general, there are many more magneto-resistive anomalies which can occur in mesoscopic systems. Many of these phenomena can be understood in the framework of Landauer–Büttiker theory, which is a description of the non-local electronic transport in a magnetic field based on dissipationless 1D conducting edge channels. In the following I will review the key ideas of quantised conductance introduced by Landauer, which were later generalised by Büttiker for multi-terminal measurements.

3.2.5 Landauer–Büttiker Formalism

As we have discussed in the beginning of this section, the conductivity (and equally resistivity) in the ballistic regime is no longer a material-specific, intrinsic quantity. Instead the non-local conductance $G = I/V$ is the property which characterises a device. In the ballistic regime, the conduction of electrons can be interpreted as the transmission probability of discrete modes between two contacts. In the simplest case of a narrow conductor (c.f. Fig. 3.12a, top) in the linear response regime, the total conductance is given by the number of modes N, which each contribute a conductance quantum $G_0 = \frac{2e^2}{h}$ to the total conductance, times the transmission probability T of the modes through the channel [66]

$$G = \frac{2e^2}{h} NT. \tag{3.6}$$

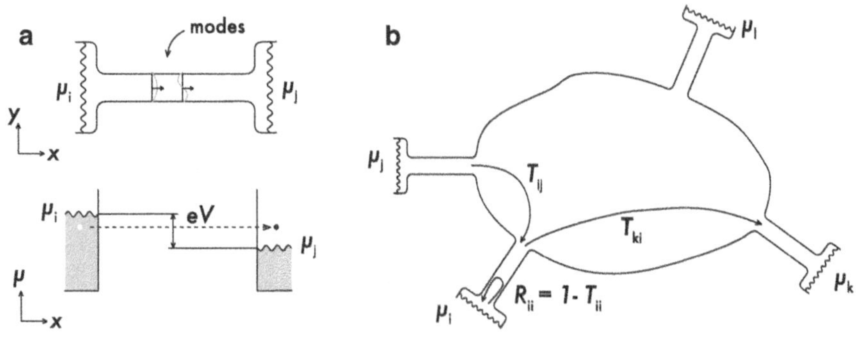

Fig. 3.12 **a** Top: Conduction modes through a ballistic channel connecting two charge reservoirs. Bottom: Electron transport across the channel leaves behind a non-equilibrium hole and creates a non-equilibrium electron. **b** Four-terminal setup according to Büttiker

Here e is the elementary charge and h is Planck's constant and we have assumed that the transmission probability of all modes is equal ($\sum_{n=1}^{N} T_n = NT$). The Landauer formula (3.6) can be rewritten in form of a resistance

$$G^{-1} = \underbrace{\frac{h}{2e^2N}}_{\text{contact resistance}} + \underbrace{\frac{h}{2e^2N}\frac{1-T}{T}}_{\text{residual scattering}}.$$

This formally separates the resistance originating in the contacts from the resistance due to residual scattering in the channel. In the absence of backscattering the transmission through the channel becomes perfect and $T \to 1$. In this case the resistance of the conductor remains finite and is generated entirely in the contacts, as the transport through the channel itself is dissipationless. This is one of the key notions of Landauer's theory. In this formalism it is useful to describe the ohmic contacts of a device as electron reservoirs at a electrochemical potential μ. By applying a voltage difference V across the channel, an imbalance in the chemical potential, $\mu_i - \mu_j = \Delta\mu = eV$, transports an electron with charge e from one side to the other (c.f. Fig. 3.12a, bottom). The finite contact resistance arrises from the non-equilibrium hole left behind at the contact i and the non-equilibrium electron arriving at contact j, which both dissipate energy by relaxing to the respective electrochemical potentials.

The description of a ballistic channel as an ideal conductor connecting two charge reservoirs was later generalised to multi-terminal systems by Büttiker [67], who realised that the current and voltage contacts should be treated on equal footing, c.f. Fig. 3.12b. The transport through a conductor of arbitrary shape with i leads can then be formulated in terms of i charge reservoirs which are connected to the conductor via narrow leads with finite transmission probability $T_{ii} = 1 - R_{ii}$. The net current I_i in contact i in the linear regime is given by

$$I_i = \frac{e}{h}\left[(N_i - R_{ii})\mu_i - \sum_{j \neq i} T_{ij}\mu_j\right]. \tag{3.7}$$

Here N_i is the number of conducting channels in the contact i, μ_i is the chemical potential of contact i, R_{ii} is the coefficient for carriers in lead i to be reflected back into lead i, and T_{ij} is the transmission coefficient from contact j to i. Based on time-reversal symmetry and current conservation, the transmission coefficients T_{ij} have the symmetry $T_{ij}(B) = T_{ji}(-B)$. As a consequence, the conductance of a four terminal system $G_{ij,kl} = \frac{V_{ij}}{I_{kl}}$ has the reciprocal relation

$$G_{ij,kl}(B) = G_{kl,ij}(-B), \tag{3.8}$$

which is known as the generalised Onsager relation. In other words the conductance (and equally resistance) of a ballistic system is symmetric under reversal of the magnetic field if the current and voltage leads are reversed simultaneously. This symmetry is the non-local analogue of the Onsager–Casimir reciprocal relation for the resistivity tensor in the diffusive regime $\rho(B) = \rho^T(-B)$, which is derived based on microscopic reversibility [68, 69].

As long as inelastic scattering events are absent and the charge reservoirs are strongly coupled to the device, the Landauer–Büttiker formalism (3.12) can successfully describe ballistic transport experiments. This includes for instance electron transport through ballistic junctions or electron focusing experiments, which I describe in detail in Sect. 3.4.

3.3 Directional Ballistic Motion in $PdCoO_2$

Although the first materials in which ballistic effects were studied were purified elemental metals and semi-metals with non-trivial, anisotropic Fermi surfaces, many of the theories modelling the observed ballistic effects were based on the simplified assumption of a circular Fermi surface. In the 1950s and 60s it was realised that the electrical conductivity may become anisotropic in the ballistic regime, reflecting the local geometry of the Fermi surface. Several models based on ellipsoidal Fermi surfaces [70–73] were then developed, which predicted anisotropic resistivity responses for thin films and for wires oriented along different crystallographic directions. Experimentally, the anisotropic size effect was measured in thin aluminium films [74, 75] and aluminium single crystal rods [76]. However, the results could not conclusively show if the observed anisotropy was related to the underlying anisotropic shape of the Fermi surface or were due to an anisotropic relaxation time of the conduction electrons. Another possible scenario envisions electrons scattering anisotropically from different crystal facets. With the emergence of semiconductor 2DEGs, which are described by an isotropic Fermi surface, these questions were no longer relevant when studying the effects of ballistic conduction electrons and

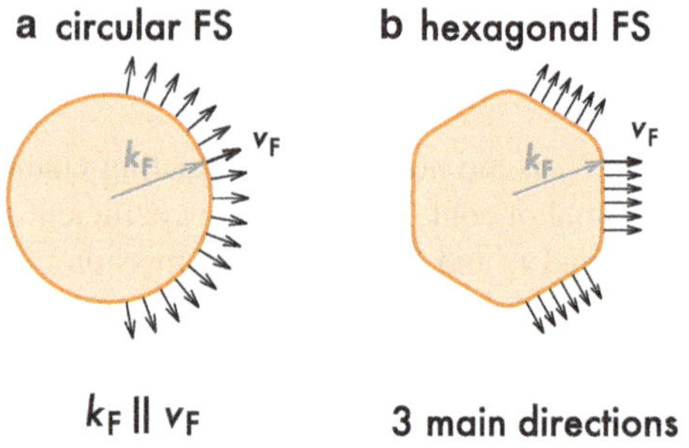

Fig. 3.13 The key concept of directional ballistics. **a** In the case of a circular Fermi surface, the Fermi velocity v_F is always parallel to the Fermi momentum k_F. **b** On a hexagonal Fermi surface, as in $PdCo_2$, v_F is restricted to 3 main directions as a result of the strongly faceted Fermi surface

remained unanswered for many years. The phenomenon of anisotropic conduction in size restricted metals has only recently seen a revival of interest as a result of the undesired resistivity increase occurring in nanoscale copper wires, which form the interconnects in semiconductor devices. Realising that the crystallographic orientation is critical in nanoscale metallic conduction, experimental and theoretical studies were performed to investigate the anisotropic effect in tungsten nanowires [77, 78]. Based on ab initio electronic structure calculations of the Fermi surface it was shown that the primary reason for observing an anisotropic size effect in epitaxial tungsten layers is due to the non-spherical Fermi surface.

Unlike for aluminium or tungsten, which have a three-dimensional electronic structure comprised of multiple bands forming the Fermi surface which complicate the analysis and interpretation of the experimental results, $PdCoO_2$ is a single band metal with a simple yet non-circular Fermi surface shape. Understanding the implications of the nearly perfectly hexagonal Fermi surface is crucial for describing the ballistic regime in $PdCoO_2$, but is also of relevance when modelling possible hydrodynamic effects that are also thought to be present in the delafossites [16, 34].

When directly comparing the Fermi surface of $PdCoO_2$ with a circular Fermi surface realised in 2DEGs we notice the following (see Fig. 3.13). The nearly free electrons (or holes) in 2DEGs, moving on a circular Fermi surface, have a Fermi velocity which always points parallel to the Fermi momentum and therefore the ballistic properties are isotropic in all directions (Fig. 3.13a). The situation in $PdCoO_2$ is very different. The Fermi velocity is restricted to mainly 3 directions and hence a forward moving electron predominantly propagates along one of these orientations (Fig. 3.13b). As a consequence the electronic system is strongly anisotropic which leads to directionally restricted ballistic motion in the ballistic regime and has some far reaching consequences which we will explore in the following section. I will start by considering symmetry arguments which will expand on the possibility of an anisotropic in-plane conductivity tensor. Thereafter I will present the results of kinetic calculations performed by Thomas Scaffidi and measurements of the in-plane longitudinal and transverse electrical transport, which are in good mutual agreement.

3.3.1 Symmetry Considerations

The conductivity tensor of a two-dimensional crystal lattice which has either 4 or 6-fold rotational symmetry is isotropic and consequently can be expressed as a scalar times the identity matrix. Therefore, when measuring the conductivity in a tetragonal (D_4) or hexagonal (D_6) crystal system, the voltage response to an applied current in a 4-point transport bar will be the same regardless of the direction in which the bar was cut. These statements will be proven in the following. In the Cartesian basis, where $\hat{e}_x$ and $\hat{e}_y$ define the plane of the 2D crystal and $\hat{e}_z$ describes the out-of-plane direction, the generic conductivity tensor C can be defined as

$$C = \begin{pmatrix} \sigma_{xx} & \sigma_{xy} \\ \sigma_{yx} & \sigma_{yy} \end{pmatrix},$$

where the z-component in a 2D system can be neglected without loss of generality. When the symmetries of the crystal system are imposed onto C, the number of independent components are reduced. We are concerned with the dihedral point groups D_n, with $n = 4, 6$, which have an n-fold rotation axes and additionally n two-fold reflection axes perpendicular to the rotation axis. The symmetry operations describing rotation and reflection can be conveniently represented in matrix form. The rotation matrix $R(\theta)$, which expresses a counterclockwise rotation through an angle θ around the out-of-plane axis is given by

$$R(\theta) = \begin{pmatrix} \cos(\theta) & -\sin(\theta) \\ \sin(\theta) & \cos(\theta) \end{pmatrix}.$$

The rotation matrix is orthogonal, i.e. it has the property $R^\top R = RR^\top = \mathbb{1}$. The rotated conductivity matrix C' is obtained by calculating $C' = R(\theta) \cdot C \cdot R(\theta)^\top$.

The matrix describing reflection about the y-axis is given by

$$S = \begin{pmatrix} -1 & 0 \\ 0 & 1 \end{pmatrix},$$

and the reflected conductivity matrix can be found by calculating $C' = S \cdot C \cdot S^\top$. Imposing this symmetry onto our generic conductivity tensor by requiring that $C' = C$ we find

$$\begin{pmatrix} \sigma_{xx} & \sigma_{xy} \\ \sigma_{yx} & \sigma_{yy} \end{pmatrix} = S \cdot \begin{pmatrix} \sigma_{xx} & \sigma_{xy} \\ \sigma_{yx} & \sigma_{yy} \end{pmatrix} \cdot S^\top = \begin{pmatrix} \sigma_{xx} & -\sigma_{xy} \\ -\sigma_{yx} & \sigma_{yy} \end{pmatrix},$$

and therefore $\sigma_{xy} = \sigma_{yx} = 0$. We conclude that the generic conductivity tensor can be simplified to exclude off-diagonal elements.

In the case of a tetragonal crystal system, the square lattice is, by definition, invariant under rotation about $\theta = \pi/2$. When imposing the C_4 symmetry on to the conductivity matrix C we find

$$\begin{pmatrix}\sigma_{xx} & \sigma_{xy}\\ \sigma_{yx} & \sigma_{yy}\end{pmatrix} = R(\theta=\pi/2)\cdot\begin{pmatrix}\sigma_{xx} & \sigma_{xy}\\ \sigma_{yx} & \sigma_{yy}\end{pmatrix}\cdot R^{\top}(\theta=\pi/2) = \begin{pmatrix}\sigma_{yy} & -\sigma_{yx}\\ -\sigma_{xy} & \sigma_{xx}\end{pmatrix}.$$

We identify that $\sigma_{xx} = \sigma_{yy} \equiv c$ and $\sigma_{xy} = -\sigma_{yx}$. With the reflection symmetry requiring $\sigma_{xy} = \sigma_{yx} = 0$ we find that $C = c \cdot \mathbb{1}$, where c is a scalar quantity. This concludes the first part of the proof, which shows that the conductivity on a square lattice can be characterised by a single scalar.

A similar formalism can be applied in the case of a triangular or hexagonal lattice. $PdCoO_2$ possesses D_6 symmetry and is therefore invariant under rotation of $\theta = \pi/3$. Again, imposing this symmetry on to the general conductivity matrix C we find

$$\begin{aligned}\begin{pmatrix}\sigma_{xx} & \sigma_{xy}\\ \sigma_{yx} & \sigma_{yy}\end{pmatrix} &= R(\theta=\pi/3)\cdot\begin{pmatrix}\sigma_{xx} & \sigma_{xy}\\ \sigma_{yx} & \sigma_{yy}\end{pmatrix}\cdot R^{\top}(\theta=\pi/3)\\ &= \frac{1}{4}\begin{pmatrix}\sigma_{xx}+\sqrt{3}(\sigma_{xy}+\sigma_{yx})+3\sigma_{yy} & \sigma_{xy}+\sqrt{3}(\sigma_{yy}-\sigma_{xx})-3\sigma_{yx}\\ \sigma_{yx}+\sqrt{3}(\sigma_{yy}-\sigma_{xx})-3\sigma_{xy} & \sigma_{yy}-\sqrt{3}(\sigma_{xy}+\sigma_{yx})+3\sigma_{xx}\end{pmatrix}.\end{aligned}$$

The system of four equations that is obtained by requiring $C' = C$ yields, after some rearranging, $\sigma_{xx} = \sigma_{yy} \equiv c$ and $\sigma_{xy} = -\sigma_{yx}$. The off-diagonal elements are again 0 as required by the reflection symmetry. We conclude that also on a triangular lattice the conductivity tensor can be written as $C = c \cdot \mathbb{1}$.

In the case of D_3 symmetry, this procedure can be repeated for $C' = R(\theta = 2\pi/3) \cdot C \cdot R^{\top}(\theta = 2\pi/3) = C$, which also restricts the conductivity matrix to a scalar $C = c \cdot \mathbb{1}$. For C_2 symmetry the reflection condition requires the off-diagonal elements to be zero, however the diagonal components do not have to be equal. So in general the conductivity matrix for a C_2 symmetric system is $\begin{pmatrix}\sigma_{xx} & 0\\ 0 & \sigma_{yy}\end{pmatrix}$.

It is then straightforward to see that in any system, in which the conductivity matrix can be expressed as a scalar times the identity matrix must have an isotropic conductivity.

Importantly, the arguments above rely on the rotational and translational symmetries of the crystal lattice. If either symmetry is broken, for instance by a lattice distortion which breaks rotational symmetry or by introducing boundaries which violate global translational symmetry, the electrical transport is no longer bound to be isotropic. For example, an orthorhombic distortion to the square lattice lowers the symmetry from D_4 to D_2, which results in generation of off-diagonal terms in the conductivity tensor when measured away from the high-symmetry directions. Using the same formalism as above we calculate the conductivity tensor C' on an orthonormal basis rotated by an angle θ for a D_2 symmetric system.

$$\begin{aligned}C' &= \begin{pmatrix}\cos\theta & -\sin\theta\\ \sin\theta & \cos\theta\end{pmatrix}\cdot\begin{pmatrix}\sigma_{xx} & 0\\ 0 & \sigma_{yy}\end{pmatrix}\cdot\begin{pmatrix}\cos\theta & \sin\theta\\ -\sin\theta & \cos\theta\end{pmatrix}\\ &= \begin{pmatrix}\sigma_{xx}\cos^2\theta+\sigma_{yy}\sin^2\theta & (\sigma_{xx}-\sigma_{yy})\cos\theta\sin\theta\\ (\sigma_{xx}-\sigma_{yy})\cos\theta\sin\theta & \sigma_{xx}\sin^2\theta+\sigma_{yy}\cos^2\theta\end{pmatrix}\end{aligned} \tag{3.9}$$

This shows how off-diagonal elements which scale with the strength of the lattice distortion $(\sigma_{xx} - \sigma_{yy})$ are generated when measuring away from $\theta = \frac{\pi}{2}n$, $n \in \mathbb{Z}$. Physically this means that the electric field $\vec{E}$ in the sample is no longer parallel to the applied current density $\vec{J}$. In the measurement of a transport bar in a Hall configuration these correspond to the generation of transverse voltages in the absence of a magnetic field. This phenomenon is used to measure conductivity anisotropy in materials which experience a tetragonal to orthorhombic phase transition [79, 80]. A transport bar cut 45° away from the principal axes will show no transverse voltage in the tetragonal phase and the maximal transverse signal $(\sigma_{xx} - \sigma_{yy})/2$ in the orthorhombic phase.

In $PdCoO_2$, the triangular lattice (C_6 symmetry) shows no indications of a lattice distortion down to lowest temperatures [15]. However, when the mean free path becomes comparable to the width of the channel, the electronic transport is influenced by the presence of the boundaries and the approximation of an infinitely extended crystal breaks down. By breaking the translational symmetry, isotropic transport is no longer required and off-diagonal elements which scale with the sample width to mean free path ratio w/l are expected. In the following I will present the results of kinetic calculations performed by Thomas Scaffidi which predict the size of this effect depending on the orientation of the transport bar with respect to the crystal lattice. The simulations are based on solving the Boltzmann transport equation for a system with a perfectly hexagonal Fermi surface and assuming diffusive scattering at the channel boundaries. Because the perfect hexagon describes the flat facets of the experimentally determined Fermi surface well, but does not capture the rounded corners, we may expect to see deviations in that direction (towards ±30°).

In the longitudinal signal (blue trace) calculated by Scaffidi in Fig. 3.14a, a resistive minimum is observed along the 0° direction which increases until it reaches a maximum at ±30°. A pictorial explanation is shown in Fig. 3.15. In the 30° direction, the Fermi surface is orientated such that 2 main electron beams are directed into the channel walls and lead to a higher resistive state. Along the 0° direction, 3 main electron beams pass through the sample. In the absence of bulk scattering, the electrons travelling parallel to the bar do not interact with the side walls, hence the overall resistance is lowered. From here on I will adapt the nomenclature "2-beam direction" and "3-beam direction" to refer to different orientations in the crystal.

The transverse signal (red trace) in Fig. 3.14 is antisymmetric with respect to the angle and has a sharp transition at 0°. This is in agreement with our symmetry considerations in the previous section which stated that ρ_{xy} is necessarily 0 along high-symmetry directions (0° and 30° in this case), but can be non-zero otherwise. The abrupt sign change around 0° can be understood by considering Fig. 3.16. The central beam of electrons running parallel to the transport bar in the 3-beam direction in Fig. 3.15B is now directed into the side walls. The statistical imbalance of electrons across the width of the sample leads to an observable voltage difference even in zero-field. From the sketch in Fig. 3.16 it also intuitively clear that this effect scales with the w/B ratio, as scattering in the bulk counteracts the electron imbalance.

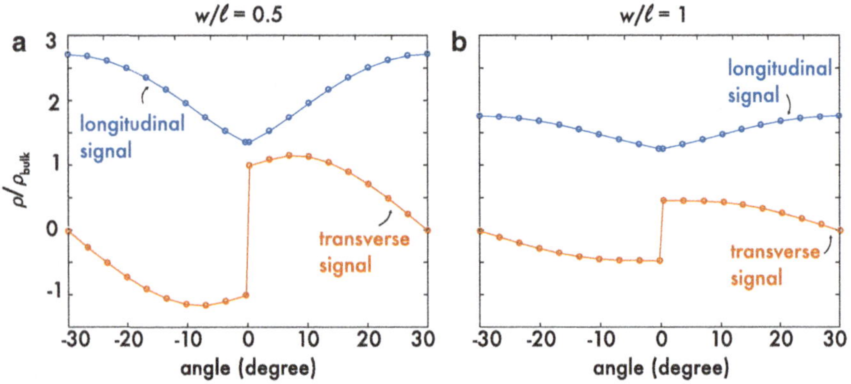

Fig. 3.14 Anisotropic longitudinal (blue) and transverse (red) resistivities at 2 K normalised to the bulk resistivity value ρ_0. In **a** the momentum relaxing mean free path is twice the width w of the channel, while in **b** they are equal. The ±30° directions point along the main crystallographic axes [100] and [010]. Accordingly the 0° direction points in-between. (Calculation and figure from T. Scaffidi.)

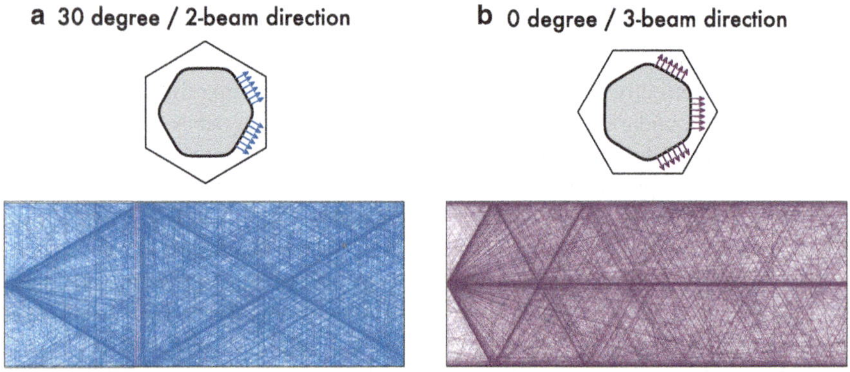

Fig. 3.15 Zero-field simulation of the electron trajectories through a transport bar in the 30° case (**a**) and in the 0° case (**b**). The orientation of the Fermi surface with respect to the bar is indicated in the top sketches

3.3.2 *In-Plane Anisotropy in* $PdCoO_2$

$PdCoO_2$ crystals were confirmed by Laue X-ray diffraction to have growth edges that are perpendicular to the crystal orientation. This permits us to directly infer the crystal orientation from the shape of the crystal with sub-degree precision. Using the FIB, four consecutive ("daisy-chained") transport bars were defined into the crystal in steps of 10° with respect to the underlying Fermi surface (Fig. 3.17). Performing

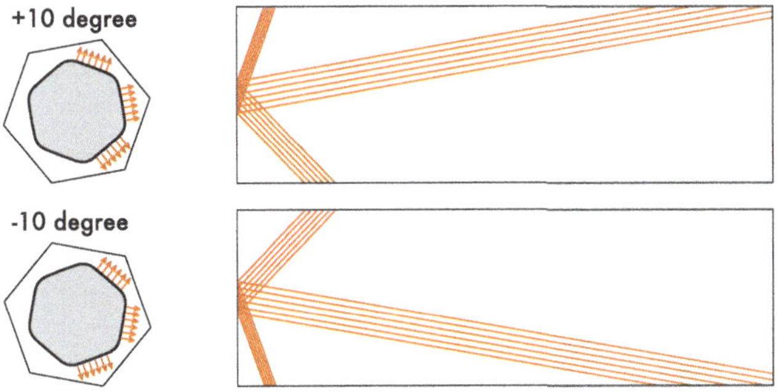

Fig. 3.16 Schematic explanation of the zero-field transverse voltage effect. The Fermi surfaces (on the left hand side) have been rotated about ±10° with respect to the transport bars. In general, when transport bars in the quasi-ballistic regime of an anisotropic system are cut along low-symmetry directions one expects an imbalance of electrons and hence an observable voltage difference to occur across the width of the sample

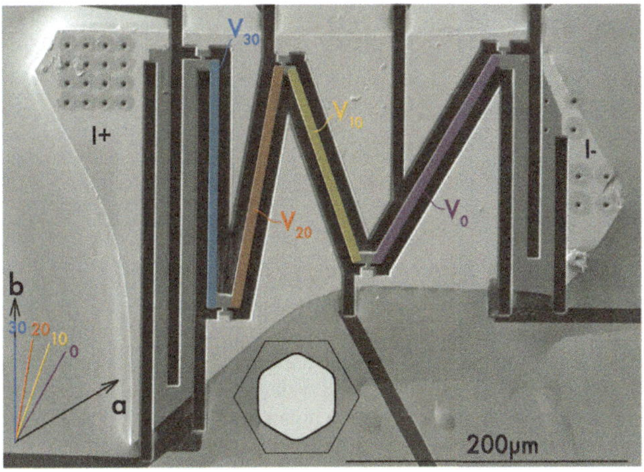

Fig. 3.17 $PdCoO_2$ device designed to measure in-plane transport anisotropy. 4 transport bars were cut in 10° steps (V_0, V_{10}, V_{20}, V_{30}), where V_{30} points along the crystallographic direction. The overlaid schematic represents the orientation of the Brillouin zone and the Fermi surface in this crystal. For fabrication details see Sect. 2.2.2

this experiment on one single crystal reduces systematic errors and ensures that the bulk defect concentration is as similar as possible across the transport bars. We also expect no difference in specularity between the sidewalls of the individual bars, as they were all cut under the same conditions and same currents.

The in-plane resistivity as a function of temperature is presented in Fig. 3.18. Down to ~50 K the transport is isotropic, as expected on a triangular lattice in the diffusive regime. At these temperatures the data are in good agreement with the values measured in bulk samples. Once the mean free path becomes comparable to the width of the sample ($w = 7\,\mu$m), however, the finite size effect increases the

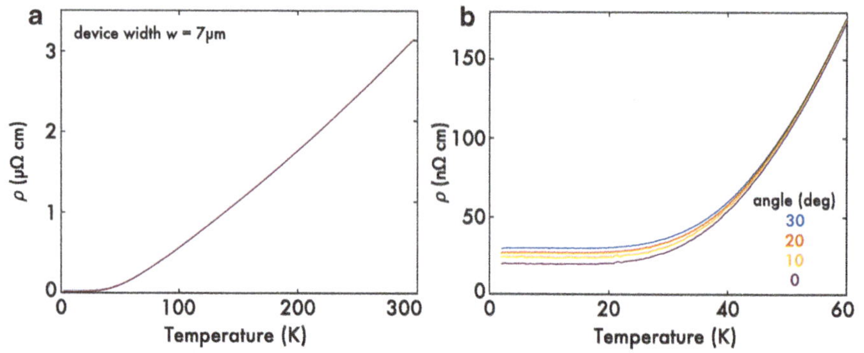

Fig. 3.18 In-plane resistivity as a function of temperature in a 7 μm wide channel. **a** The transport above 50 K is isotropic while below 50 K **b** there is an up to 50% difference in the in-plane transport

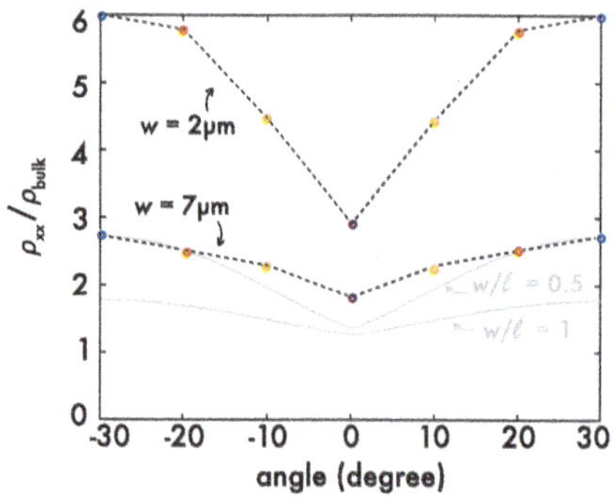

Fig. 3.19 Comparison of the measured in-plane anisotropy with the calculated resistivities. The data were obtained from one crystal device with width $w = 7\,\mu m$ which was later thinned down to $w = 2\,\mu m$. The coloured circles are the data points and the dashed line is a guide to the eye connecting them. The grey lines are Scaffidi's simulations, which were introduced in Fig. 3.14

resistivity value by roughly a factor 2 due to increased boundary scattering. Moreover, the curves split up and show an anisotropy of up to 50% as a result of the non-uniform Fermi velocity distribution. The data qualitatively agree with Scaffidi's calculations, confirming a resistivity minimum along the 0° direction which steadily increases and becomes maximal along the 30° direction.

A quantitative comparison is presented in Fig. 3.19. As the bulk resistivity value ρ_{bulk} was never directly measured in this particular sample, we instead estimate the value by considering the resistivity at high magnetic fields, which is independent of the sample width w. At 10 T (c.f. Fig. 3.20) we find $\rho_0 = 11.24$ nΩcm. From Nandi's work [16], we can estimate the magneto-resistance ($\Delta\rho/\rho_{bulk}$) to increase by about 12% from 0 to 10 T at 2 K, such that our final estimate for the bulk resistivity is about $\rho_{bulk} = 10$ nΩcm. The measurements are roughly consistent with a momentum relaxing mean free path of $l_{mr} = 2w = 14\,\mu m$. The deviations around 0° might stem

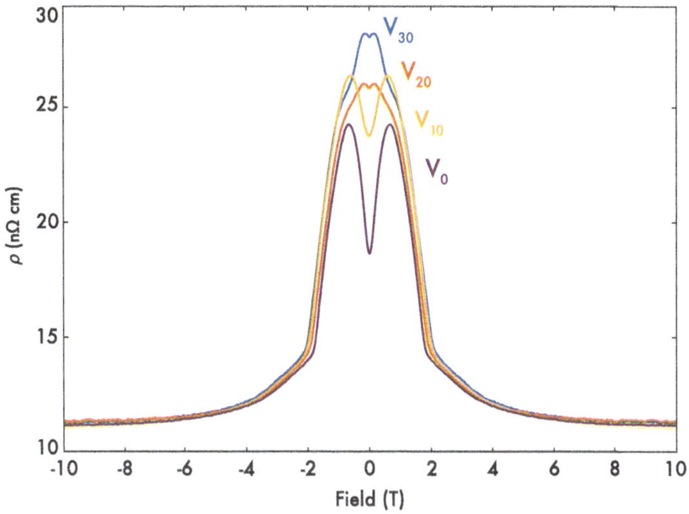

Fig. 3.20 In-plane resistivity as a function of magnetic field at 2 K (symmetrised data)

from an error in the estimation of ρ_{bulk} or from Fermi surface details, such as warping along the flat sections, which are not accounted for in the simulations. The crystal device in Fig. 3.17 was further thinned down to 2 μm. The result highlights the strong w/l dependency of the finite size effect, which also increases the anisotropy V_{30}/V_0 to roughly 200%.

The effects of the in-plane anisotropy become even clearer when investigating the magneto-transport in a size-restricted bar, shown in Fig. 3.20. Following the discussion in Fig. 3.11, we recognise that the magnetic field scale can be divided into two regimes. Beyond roughly ± 2 T the cyclotron diameter is smaller than the device width ($2r_c < w$). In this regime isotropic in-plane transport is recovered because the cyclotron radius is the smallest length scale in the system and the local translational symmetry is restored. Inside the ± 2 T region we discover the consequences of a classical size effect, which leads to an anisotropic resistivity enhancement. Along the 3-beam direction (V_0, purple), we find the familiar behaviour of a positive magneto-resistance leading to a peak, which is followed by a region of negative magneto-resistance. Along the 2-beam direction (V_{30}, blue), however, there is an overall peak rather than a dip in zero-field. This is a consequence of the two main beams of electrons in Fig. 3.15, which are already maximally directed into the boundary at zero-field. The application of a magnetic field will cause the electrons to statistically hit the boundaries less often and hence the overall magneto-resistance is decreasing. The tiny dip at zero-field is likely a result of the rounded corners of the nearly perfectly hexagonal Fermi surface. The other cases, V_{10} and V_{20}, can be understood as cases intermediate between the 2-beam and 3-beam directions.

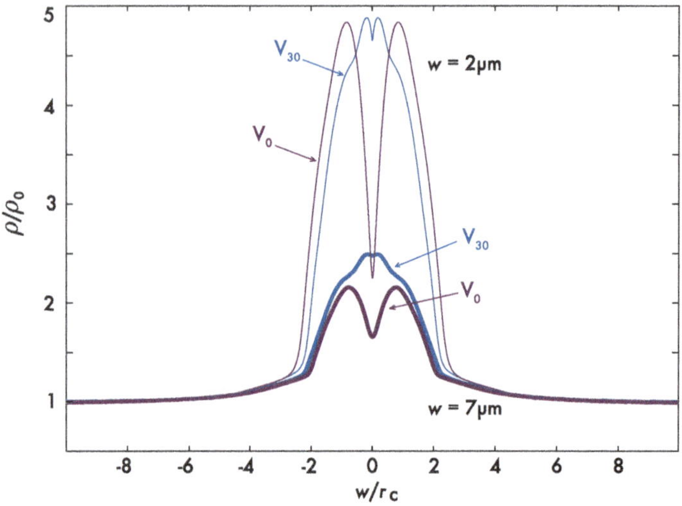

Fig. 3.21 Comparison of the anisotropic magneto-transport between a 7 μm (thick lines) and 2 μm wide (thin lines) device. For clarity only V_0 and V_{30} are shown. The axes have been converted to the dimensionless quantities w/l and ρ/ρ_0, where ρ_0 is the resistivity value taken at 16T

As an additional test, we use the FIB to reduce the channel width from 7 μm to 2 μm which more than triples the w/l ratio. Since this experiment is performed on the very same crystal and hence only w is altered, while l is left unaltered, we can exclude any effects which would stem from varying crystal quality. In Fig. 3.21 the data are plotted as a function of the dimensionless parameters w/r_c and ρ/ρ_0, where ρ_0 is the resistivity value taken at 16 T. In the 2 μm wide sample, the anisotropy in 0 field has increased to 200 %. The features in the low field region have more than doubled in magnitude and become more pronounced, but the overall qualitative behaviour has remained the same.

3.3.3 Zero-Field Transverse Voltage in $PdCoO_2$

We now set out to investigate the zero-field transverse voltage (ZFTV) in an anisotropic system as theoretically predicted by Scaffidi in Fig. 3.14. We expect this effect to be sensitive to the ratio of w/l which quantifies the degree of translational symmetry breaking, similar to ZFTV measurements in orthorhombic systems which are proportional to the magnitude of lattice distortion $(a - b)$.

The device shown in Fig. 3.22 was specifically designed to probe zero-field transverse voltages. This is a challenging transport experiment for multiple reasons:

- The absolute signal sizes are small. The transverse resistance is given by the resistivity of the sample divided by the thickness t of the sample. Using the calculations

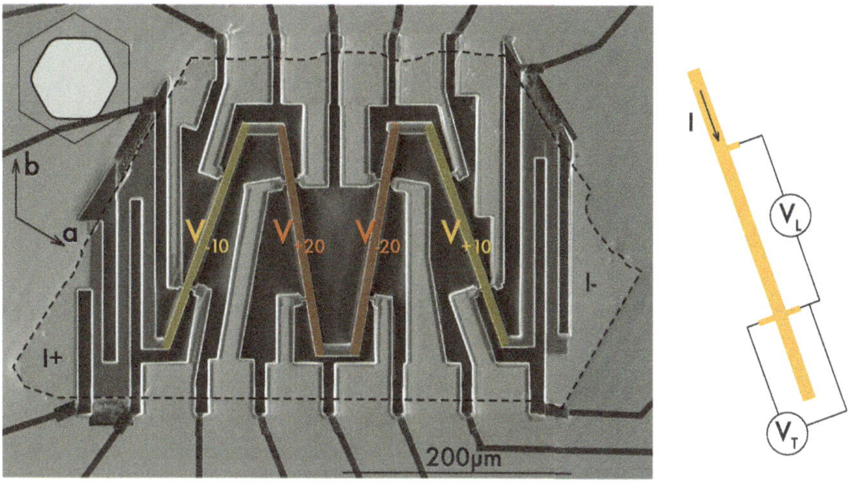

Fig. 3.22 $PdCoO_2$ device designed for measuring a zero-field transverse voltage contribution along low-symmetry directions. The outline of the crystal is traced by a dashed line. The current path through the device from $+I$ to $-I$ incorporates meandered sections in the beginning and at the end to homogenise the current, which injected via gold top contacts. The device consists of 4 transport bars cut along ±10 (yellow) and ±20 (red) direction each with 3 voltage contacts to probe the longitudinal and transverse voltage simultaneously, c.f. schematic on the right. The crystal is 8 μm high and each of the transport bars is about 6 μm wide and 100 μm long. The overlaid schematic in the top left corner shows the orientation of the Brillouin zone and the Fermi surface with respect to this crystal

in Fig. 3.22a for an estimate and choosing a high excitations current ($I = 5\,\mathrm{mA}$), we expect to measure a voltage difference

$$\Delta V = R \cdot I = \frac{\rho}{t} \cdot I = \frac{0.5\rho_{bulk}}{t} \cdot I = \frac{0.5 \cdot 8\,\mathrm{n\Omega cm}}{8\,\mu\mathrm{m}} \cdot 5\,\mathrm{mA} = 25\,\mathrm{nV}.$$

This of course can be improved by thinning the crystal and thereby reducing its thickness or by reducing the channel width.

- The signal from any transverse contact pair will always contain a longitudinal component, because the contacts are never perfectly opposed, even when structured with the FIB. In a Hall measurement the transverse component ρ_{xy} can be isolated from the signal by exploiting the fact that it transforms antisymmetrically in field ($\rho_{xy}(B) = -\rho_{xy}(-B)$), while the longitudinal component is symmetric ($\rho_{xy}(B) = \rho_{xy}(-B)$). For a zero-field measurement this procedure is not possible. Instead, the longitudinal component has to be measured simultaneously and subtracted from the transverse signal by fitting the curves at room temperature.
- While in a typical diffusive transport experiment the width of the sample only influences the resistance linearly ($\rho = R\frac{w \cdot t}{l}$), this is not the case in the ballistic regime. Hence the transport bars must have an equal width over their entire length

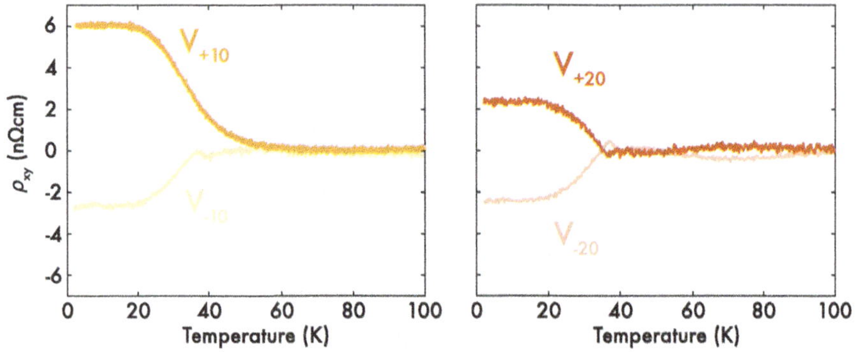

Fig. 3.23 Zero-field transverse voltage measurements as a function of temperature. 4 transport bars (c.f. Fig. 3.22) were measured along the low symmetry directions ±10° and ±20°. The sample width w was 6 μm except for V_{10}, where it was 8 μm

of 100 μm and should also be equal in length to each other for a quantitative comparison of the results.

The data in Fig. 3.23 clearly show the onset of a ZFTV signal around 40 K. As theoretically predicted, the results are antisymmetric with respect to the angle they were measured at. The reason for the asymmetry between V_{+10} and $-V_{-10}$ lies in the varying widths w of the samples and hence different w/l ratios. One curiosity that strikes the eye is the unusual bump at 40 K, which can be seen in many of the curves. This is unexpected as the mean free path increases smoothly between 50 and 20 K. One possible explanation could lie in the finite width of the transverse voltage contacts themselves which locally increase the sample width right at the point where the voltage is measured. A future measurement of this sample with a thinned down width and slimmer voltage contacts should hopefully bring clarity. Moreover the absence of ZFTV along the 0° and 30° direction will be tested.

Comparing the measured transverse resistivity to the predictions by Scaffidi (c.f. Fig. 3.24), we find that this particular 6 μm wide sample agrees well with the predictions if it has a mean free path somewhere between 6 μm and 12 μm. In agreement with the calculations, the ZFTV at ±20° is smaller than that at ±10°. To precisely map out the overall angle dependent shape, a sample with more transport bars orientated along various directions could be fabricated.

In conclusion, in this section I have explored the directional ballistic effects occurring in $PdCoO_2$, which are also expected to be present in the closely related delafossite compounds $PtCoO_2$ and $PdRhO_2$. In the ballistic regime, the electronic motion is restricted to mainly three directions due to the underlying nearly perfectly hexagonal Fermi surface. The consequence of the nearly hexagonal Fermi surface shape is reflected in the anisotropic size-effect which occurs in restricted channel experiments and is strongly dependent on the w/l ratio. The effect is particularly pronounced in the angle-dependent magneto-resistance. The qualitative shape, in absence but also in presence of a magnetic field can be well understood in a simple semi-classical

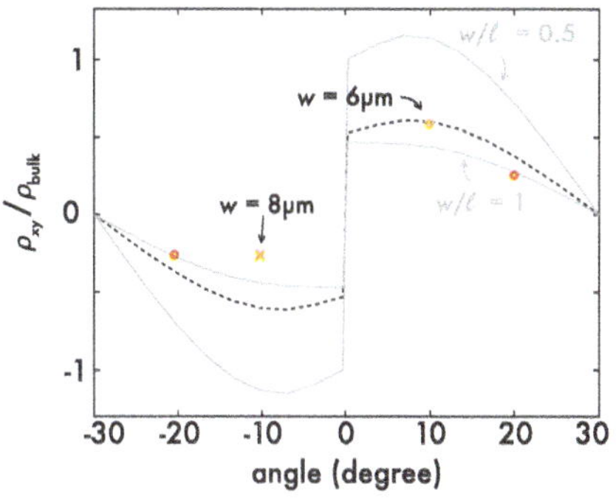

Fig. 3.24 Comparison of the predicted zero-field transverse voltage (in grey) to the measured values of a 6 μm wide sample (coloured circles). The data point from the 8 μm wide sample is indicated with a cross. The dashed line is a guide to the eye. A value of $\rho_{bulk} = 10\,\mathrm{n\Omega cm}$ was assumed

picture of "billiard-ball" electrons moving on a hexagonal Fermi surface. The broken translational symmetry is further reflected in the occurrence of the theoretically predicted zero-field transverse voltages along low-symmetry directions, which have been experimentally confirmed. These insights give us an improved microscopic understanding of the directional ballistic effects in $PdCoO_2$, which lay the basis for exploring potential hydrodynamic effects. In particular, these results establish that a simple circular Fermi surface model is not sufficient and should be replaced by a hexagonal Fermi surface. From an application point of view, the strongly faceted Fermi surfaces bear advantages compared to the standard round Fermi surface. In a ballistic device both the resistive anisotropy (0° vs. 30°) as well as the enhanced electron collimation can be exploited. The full potential of the induced anisotropy has not yet been experimentally explored; it is simply not known how big it could be made if channels a few hundred nm wide could be aligned with the relevant crystal axis to better than 0.1 degree precision. I will elaborate more on the unusual electron collimation, which eventually leads to the so called super-geometric focusing effect in the next section about transverse electron focusing.

3.4 Transverse Electron Focusing

Transverse electron focusing (TEF) experiments, otherwise known as transverse magnetic focusing (TMF) or magnetic electron focusing, probe the conduction electrons (or holes) of a ballistic system under the influence of a magnetic field $\vec{B}$. The field couples to the charge carriers via the Lorentz force $\vec{F_L} = e(\vec{E} + \vec{v_F} \times \vec{B})$, which directs them onto cyclotron orbits with a radius $r_c \propto 1/B$ in a plane perpendicular to the magnetic field. The ballistic electrons, which propagate around the edge of the sample in skipping orbits, can be probed by two narrow contacts which are placed side-by-side along the sample edge. When the magnetic field is tuned such that the cyclotron diameter $2r_c$ is equal to the spacing L between the narrow contacts, the electrons can directly skip from one contact to the next, under the condition that their mean free path l is larger or comparable to the contact separation. This effect is called transverse electron focusing and is characterised by a distinctive peak in the voltage signal as a function of the applied magnetic field. The peak position depends on the nozzle spacing L and on the Fermi wave vector k_F such that, in principle, the shape of the Fermi surface can be investigated [81]. As the Lorentz force depends on the charge of the quasi-particle, TEF is a powerful probe to detect and study composite fermions [82, 83]. In recent years, TEF experiments were also performed to investigate the Fermi surfaces in 2DEGs in which spin-orbit coupling combined with inversion symmetry breaking leads to spin-split peaks [84, 85].

One key advantage of TEF over a transport experiment in a traditional bar shaped sample is the possibility of identifying a subset of electrons which have directly propagated from one nozzle to another without having interacted with the boundary. This is of particular importance in ballistic systems, where boundary collisions play a significant or dominant role, but are hard to treat theoretically because microscopic details matter, such as the surface roughness and specularity. Having isolated the electrons that are not influenced by the boundary, the bulk mean free path l can be extracted by varying the separation L between the nozzles and thus increasing the possibility of bulk scattering events [86]. This is one of the aims of studying TEF in $PdCoO_2$.

TEF experiments were first carried out in elemental metals (Bi [87], Sb [88], W [89], Cu [89], Ag [90], Zn [91], Al [92]) in the 1970s and 80s, where purification techniques lead to extremely high quality single crystals with mean free paths reaching up to several millimetres [36]. The main interest at the time was the determination of the reflection coefficient for various surface boundaries, by comparing the voltage signal contributions from electrons which had interacted with the boundaries exactly once to those which did not collide with the sample edge. Sharp needle-like micro-contacts, so called Sharvin probes, were pressed into the crystal facets and much care had to be taken not to damage the surface [91]. Numerous studies were published and even compared with surface measurements by scanning-tunneling-microscopes (STM) [93]. However many of these analyses remain questionable due to the simplifying assumptions made and possible surface damage from the needle probes themselves.

In the late 1980s and 1990s the rise of high quality 2DEGs meant TEF could be studied in a planar geometry [94]. The improved spatial resolution offered by lithographical fabrication technologies led to well-controlled experiments. Moreover, the circular Fermi surface simplified the theoretical treatment and led to a good understanding and description of the ballistic electronic system in semiconductor 2DEGs. Most recently, when disorder scattering in graphene was significantly reduced by encapsulating it in hexagonal Boron-Nitride (hBN) in the late 2000s [95], the mean free path became long enough for TEF to be detected even at room temperature [96, 97]. Unlike semiconductor 2DEGs, the 2D electron system in graphene hosts chiral electrons [98]. These can give rise to unique electron-optic effects, such as negative refraction across a p-n junction [99], which add new opportunities for solid state electronic optics.

The two key differences between TEF in a two-dimensional semiconducting system and the delafossite metal $PdCoO_2$ are

- the Fermi surface shape: The circular Fermi surface of a 2DEG leads to an isotropic in-plane transport and TEF spectrum, while the nearly hexagonal Fermi surface in $PdCoO_2$ gives rise to a direction dependent, non-trivial TEF signal.
- the Fermi wavelength and therefore the surface specularity: The small Fermi wave vector k_F in 2DEG systems gives rise to a comparatively large Fermi wavelength $\lambda_F \approx 26$ nm in GaAs, for example, compared to $\lambda_F \approx 0.65$ nm in $PdCoO_2$ (c.f. Table 3.1). Comparing these values to the average surface roughness, which is estimated to be on the order of nanometers, we expect the surface specularity to be much greater in the case of a 2DEG compared to the delafossite system.

3.4.1 The Basic TEF Setup

A classical TEF setup is sketched in Fig. 3.25. Two very narrow (~250 nm) constrictions resembling "nozzles" are carved into the single crystal with the FIB. The distance L between the two nozzles should be smaller or comparable to the mean free path l. Two further large electrodes are placed far away from the nozzles and complete the basic device setup. A current is sourced from one of the nozzles to a large electrode, while the voltage difference between a second nozzle and the other large electrode is measured. Applying a uniform, out-of-plane magnetic field B forces

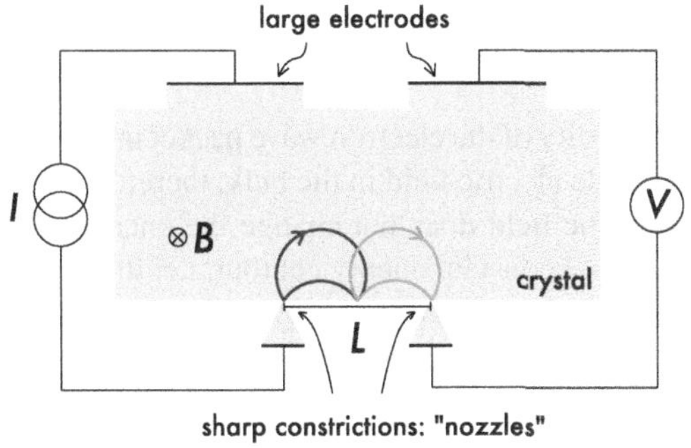

Fig. 3.25 Typical transverse electron focusing (TEF) measurement setup. The grey shaded region represents the single crystal device. Two narrow constrictions resembling "nozzles" are fabricated on the same edge of the device and two further large electrodes are created on the opposite sample side. A current is driven between one of the nozzles and one large electrode, while the voltage is measured between the remaining second nozzle and large electrode. When the distance L between the nozzles is shorter than or comparable to the mean free path l, ballistic electron effects are expected to contribute to the voltage response

the electrons onto orbits with cyclotron radius r_c, which decreases with increasing magnetic field:

$$r_c = \frac{\hbar k_F}{eB}. \tag{3.10}$$

Here $\hbar$ is the reduced Planck constant, k_F is the Fermi wave vector, e is the elemental charge and B is the magnetic field strength. If the field is tuned such that the cyclotron diameter $2r_c$ is equal to the nozzle spacing L, a distinctive voltage peak can be detected. Upon increasing the magnetic field, the focusing condition is met again when the nozzle spacing is an integer multiple n of the cyclotron diameter, $L = n \cdot 2r_c$. With that we expect a series of peaks, spaced linearly in field at positions

$$B_n = \frac{2n\hbar k_F}{eL}. \tag{3.11}$$

3.4.2 TEF of a Nearly Hexagonal Fermi Surface

So far we have implicitly assumed a system with an isotropic Fermi surface. This simplification breaks down in the case of many materials, including $PdCoO_2$. Electrons orbiting non-circular Fermi surfaces in k-space will follow a non-circular path in real space accordingly. In fact, any real space orbit is simply determined by the k-space orbit which is scaled and rotated by 90° around the magnetic field axis. This can be seen from considering the equation of motion of an electron in a solid, which is subject to a magnetic field $\vec{B}$. By equating the Lorentz force to the derivative of the crystal momentum, we arrive at

$$e(\dot{\vec{r}} \times \vec{B}) = \hbar\dot{\vec{k}}, \tag{3.12}$$

where $\dot{\vec{r}}$ is the group velocity of the electron wave packet in real space. In the ballistic regime, there is negligible electric field in the bulk, therefore I have set $\vec{E}$ to 0. First we note that the magnetic field does not change the energy of a charged particle, such that it will move on a constant energy contour, i.e. the Fermi surface. This can be checked by calculating the change in energy ϵ of the particle

$$\frac{\partial\epsilon}{\partial t} = \frac{\partial\epsilon}{\partial k}\frac{\partial k}{\partial t} = \hbar\dot{\vec{r}} \cdot \frac{e}{\hbar}(\dot{\vec{r}} \times \vec{B}) = 0.$$

Next, without loss of generality, we choose to apply the magnetic field along the $\hat{z}$ direction, $\vec{B} = (0, 0, B)$. By performing the time integration we can rewrite the expression in Eq. (3.12) as

$$r_y(t) - r_y(0) = \frac{\hbar}{eB}[k_x(t) - k_x(0)]$$

$$-r_x(t) + r_x(0) = \frac{\hbar}{eB}[k_y(t) - k_y(0)]$$

From this it is clear that the real space orbit $(r_x(t), r_y(t))$ is given by the k-space orbit $(k_x(t), k_y(t))$ rotated by 90° and scaled by a factor $\frac{\hbar}{eB}$. For this reason the ballistic electrons in $PdCoO_2$ trace a hexagonal path (c.f. Fig. 3.4) in a magnetic field. The consequences of having hexagonal rather than circular cyclotron orbits are illustrated in Fig. 3.26, which displays TEF simulations of both.

Simulations of semi-classical ballistic particles propagating on an arbitrarily shaped cyclotron orbit in a solid can be numerically implemented rather easily. We treat the electron wave as point particles with energy E_F and velocity v_F, that propagate on a path determined by the underlying Fermi surface shape. Electron-electron and electron-impurity scattering have been neglected. For the sake of simplicity, the particles are ejected isotropically in all directions from the nozzle. The sample edges are modelled as flat boundaries without surface roughness, which have the property of scattering an electron specularly with a probability p and diffusively with probability $(1 - p)$. Here we have set $p = 1$.

Panel a of Fig. 3.26 displays the simulated electron trajectories of a system with a circular Fermi surface. The grey triangle represents the ejection nozzle from which the electrons are emitted at an angle θ, where θ is defined to be 0° when the electrons are emitted perpendicular from the sample edge. The point of impact x of the electron with the sample boundary directly depends on the ejection angle θ. Simple trigonometry determines the impact position to be $x = 2r_c \cos(\theta)$. The electrons which are emitted perpendicular to the sample edge propagate the furthest $x = 2r_c$ and the impact points of electrons which are emitted at an angle $\pm\theta$ are closer to the nozzle the larger θ is. The crucial point in understanding the phenomenon of electron focusing is to realise that the density of electrons which arrive at $x = 2r_c$ diverges as $\theta \to 0$. In other words, only the electrons which are ejected perpendicular or nearly perpendicular from the sample edge are relevant for the transverse electron focusing peak. This is illustrated in panel Fig. 3.26c, which shows the histogram of the simulated points of impact of 50000 particles moving on circular cyclotron orbits. The electrons propagating inside the purple shaded region in panel a form the focusing peak in panel c below. Of these electrons, the ones which are specularly reflected form the higher harmonic focusing peaks at $x = n \cdot 2r_c$. All other electrons only contribute to the background.

In general the Fermi surface shape and orientation shape influences the TEF peak shape. In systems with Fermi surfaces with faceted edges, like the flat sides of hexagon in $PdCoO_2$, this can lead to a strong enhancement of the focusing peak. For the situation displayed in panel c of Fig. 3.26, there is a macroscopic number of electrons exiting the nozzle at $\theta = 0$. All of these particles will naturally impact the sample edge at the same point, leading to a geometric enhancement of the focusing peak, so called super-geometric focusing (or 'sfocusing' for short). By comparing the

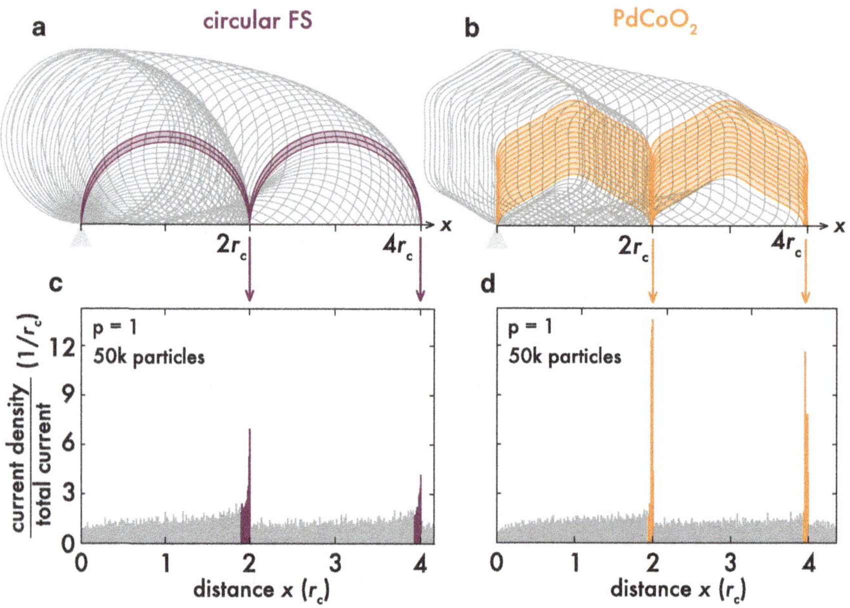

Fig. 3.26 Simulation of cyclotron orbits in real space in the case of a **a** nearly hexagonal and **b** circular Fermi surface. The grey shaded triangles represent the current ejecting nozzles. The ejection angle of an electron will determine the point of impact with the device boundary. Electrons ejected perpendicular or nearly perpendicular from the sample edge are focused onto the same point on the boundary (shaded in orange or purple). This leads to the characteristic focusing peak in panels **c** and **d**, which show a histogram of the electron distance x the electrons have moved along the sample edge. In the case of $PdCoO_2$, the flat sides of the Fermi surface boost this effect leading to a greatly enhanced focusing peak. This simulation was performed with 50000 electrons and setting the boundary specularity to be perfect ($p = 1$)

simulated TEF spectra in C and D we expect the sfocusing amplitude to be roughly twice the TEF amplitude of an isotropic system if broadening effects by finite nozzle widths are neglected.

The importance of the orientation and detailed shape of the Fermi surface is emphasised in Fig. 3.27. We consider the high-symmetry directions where the underlying hexagonal Fermi surface is orientated such that either a corner ('2-beam direction') or flat side ('3-beam direction') is aligned with the nozzle, c.f. inset in panels a and b. This leads to two distinct focusing spectra as we will see in the following. In the case of the 3-beam direction, there are mainly three electron jets ejected from the nozzle. Recalling that only the electrons which are emitted close to $\theta = 0$ are of relevance for the peaks in the TEF spectrum, we expect an enhanced amplitude of the focusing peak as a macroscopic number of electrons fulfil this condition. This is confirmed by the results of the simulation in panel d, which show a large peak in the TEF spectrum. The super geometrical focusing effect is even stronger in a perfect hexagon as indicated by the resulting sharp TEF peak (in grey). On the other hand,

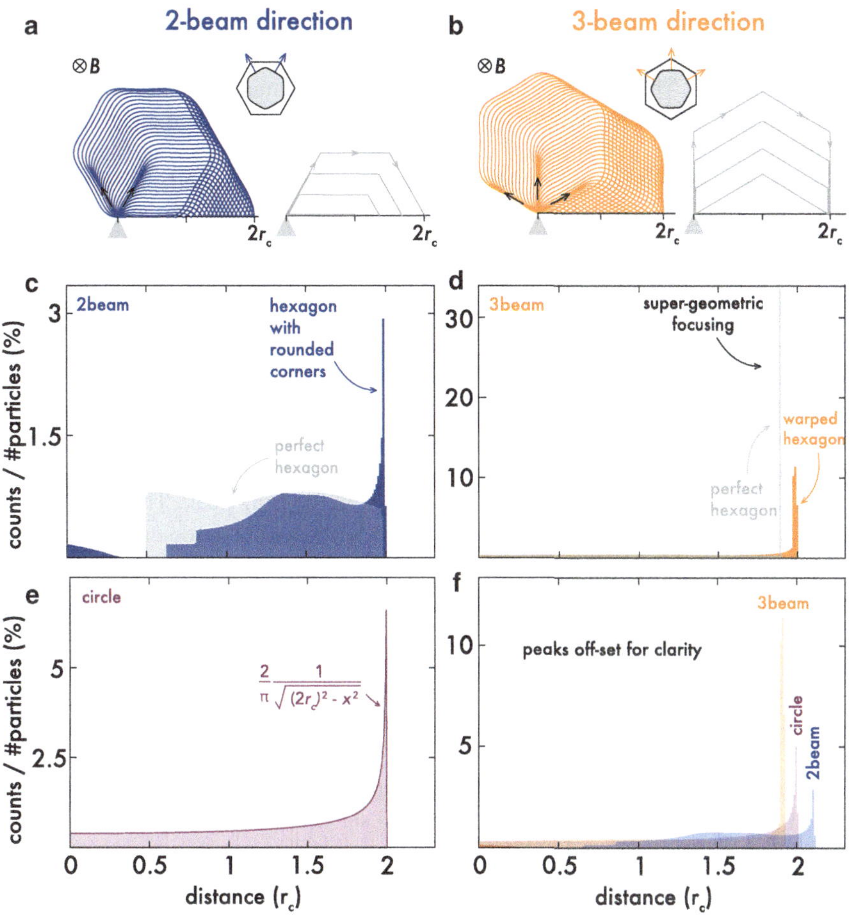

Fig. 3.27 Focusing peak shape comparison for different Fermi surface shapes and orientations. In **a** and **b** the real space electron trajectories (in colour) depending on the Fermi surface orientation (inset) are compared. The real space trajectories based on perfect hexagon emphasise the difference between both directions. **c** and **d** show a histogram of the points of impact with the sample boundary for the perfect hexagon and the nearly hexagonal Fermi surface in $PdCoO_2$. The Fermi surfaces are orientated such that either the corner (2-beam direction) or the flat sides (3-beam direction) are parallel to the electron ejection nozzle. For a circular Fermi surface **e**, the simulations match the analytically derived expression perfectly, which diverges as $x \rightarrow 2r_c$. **f** shows the difference between the 3-beam (sfocusing) and 2-beam (nocusing) orientation on the same scale

in a perfect hexagon oriented in the 2-beam direction, there are zero electrons which exit the nozzle at $\theta = 0$ and hence we expect to find no focusing (a direction we call 'nocusing'). This is verified by the results of the simulation in panel c. Rather than a pronounced focusing peak, there is a broad distribution of electron impact points. We note that in this particular case there is a minimal distance that the electron must travel before impacting the boundary, which corresponds to the length of one edge of the hexagonal trajectory. For the imperfect hexagonal Fermi surface in $PdCoO_2$, however, the simulations reveal a focusing peak nonetheless. This is due to the rounded corners of the hexagon which allow electrons to exist the nozzle at angles close to $\theta = 0$. The distinctive shape of the background which forms a side peak is determined by details in the warping of the flat regions of the Fermi surface. For Fermi surface shapes which are convex it is straightforward to derive the peak shape analytically. The calculation for a circular Fermi surface is performed in the next paragraph. The resulting function (purple curve) is displayed in panel e and traces the simulated data perfectly (purple shaded region). The divergence of the function as $x \to 2r_c$ gives rise to the observed focusing peak. The side-by-side comparison of the focusing peaks in panel f reveals an amplitude enhancement of roughly 50% for the sfocusing orientation and an equal reduction for the nocusing direction compared to an isotropic Fermi surface. In these simulations the angular distribution of electrons ejected from the nozzle was assumed to be isotropic. In reality the nozzle typically generates a directed beam of electrons, which depends on the shape of the nozzles, the Fermi surface shape and boundary specularity. For a circular Fermi surface the electron injection distribution is often approximated to vary as $\cos(\theta)$ [100].

Finally I will analytically derive the TEF spectrum in the case of (i) an isotropic and (ii) a directed injection in a circular Fermi surface system. Employing classical probability theory, let X and Θ be continuous variables, where $X = g(\Theta)$ describes the distribution of impact points x depending on the ejection angle θ. Further, let $f_\Theta(\theta)$ be the probability density function (PDF) describing probability of θ falling within the infinitesimal interval $[\theta, \theta{+}\mathrm{d}\theta]$. We are looking for the TEF peak shape which is given by the PDF of X, $f_X(x)$, which describes the probability of x falling into an interval $[x, x{+}\mathrm{d}x]$. The sought for PDF transformation is given by

$$f_X(x) = f_\Theta(g^{-1}(x)) \cdot \left|\frac{d}{dx} g^{-1}(x)\right|, \tag{3.13}$$

where we have $x = g(\theta) = 2r_c \cos(\theta)$ for a circular Fermi surface.

(i) An isotropically injected beam of electrons is described by a uniform density in the interval $[-\frac{\pi}{2}, \frac{\pi}{2}]$, such that the PDF is given by

$$f_\Theta(\theta) = \begin{cases} \frac{1}{\pi}, & \text{if } \theta \in [-\frac{\pi}{2}, \frac{\pi}{2}]. \\ 0, & \text{otherwise.} \end{cases}$$

Requiring the PDF to be normalised, $\int_{-\infty}^{\infty} f_X(x) = 1$, we find $f_X(x) = \frac{2}{\pi}\frac{1}{\sqrt{4r_c^2 - x^2}}$. This corresponds to the curve plotted in Fig. 3.27e.

(ii) Similar for a beam that is injected in the form of a cosine, the probability density function of θ is given by

$$f_\Theta(\theta) = \begin{cases} \frac{1}{2}\cos(\theta), & \text{if } \theta \in [-\frac{\pi}{2}, \frac{\pi}{2}]. \\ 0, & \text{otherwise.} \end{cases}$$

With the transformation (3.13), the TEF spectrum can be calculated, $f_X(x) = \frac{x}{2r_c}\frac{1}{\sqrt{4r_c^2 - x^2}}$.

In general if the nozzle injects the electron beam in the form of a cosine, this will cause an enhancement the focusing peak as more electrons are ejected close to $\theta = 0$. In the case of $PdCoO_2$ it is not clear if the assumption of a cosine distribution, which was derived for a circular Fermi surface, is valid. For instance in the 2-beam direction one might expect a distribution which is peaked at $\theta = \pm 30°$, where the main electron beams are ejected from the nozzle.

3.4.3 TEF in $PdCoO_2$

Modern lithographical techniques which are well suited for processing two dimensional systems are used for fabricating TEF devices in 2DEGs and graphene. These methods, however, can often not be used for arbitrarily shaped three-dimensional crystals. Instead, a convenient and straightforward way of fabricating a transverse electron focusing device from a single crystal is by using focused ion beam (FIB). A step-by-step guide of the fabrication process is described in Sect. 2.2.2; here I will focus only on the key features of $PdCoO_2$ TEF structures. In general, the requirements for a TEF device are

- non-invasive, ohmic contacts, i.e. contacts which do not damage the crystal bulk or its surface during fabrication or operation,
- narrow constrictions, $\ll l$, which form the nozzles used to probe the electron skipping orbits and
- closely spaced nozzles, smaller or comparable to l.

These demands can be met by using a FIB fabrication approach. To start, a $PdCoO_2$ single crystal is epoxied onto a substrate and top-contacted with a thin layer of gold (c.f. Fig. 3.28a). This forms the ohmic contacts to the device. The single crystals grow as roughly 10–20 μm thin platelets with the c-axis pointing of the plane, such that the palladium planes, which host the quasi-two-dimensional ballistic electron system, are always oriented parallel to the substrate. In attempt to sustain a homogenous current flow along the planes, through-holes were FIB-milled through the crystal with the idea of improving the interlayer coupling (c.f. Sect. 2.2.2). The central region of the

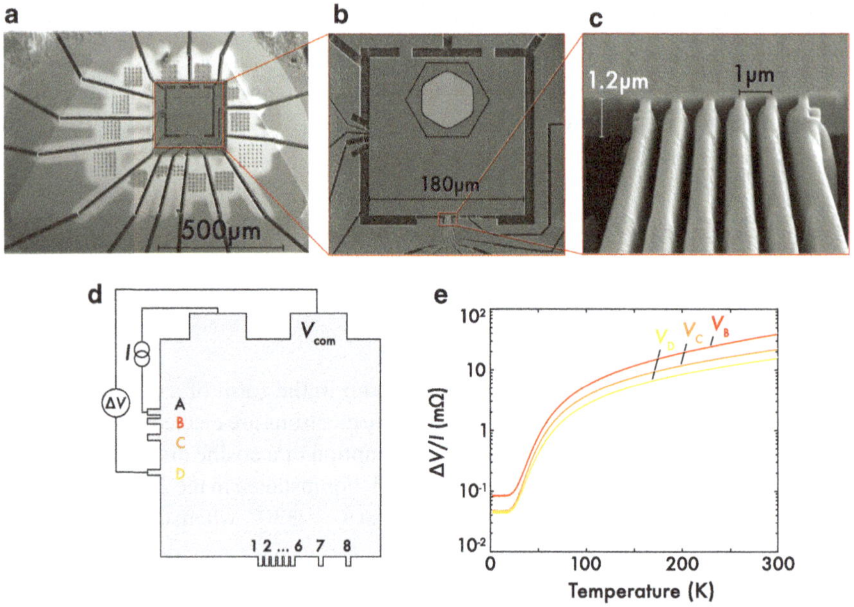

Fig. 3.28 TEF device and sample characterisation. **a** A $PdCoO_2$ crystal which was structured into a TEF geometry. Gold top contacts connect to large current and voltage electrodes, which were carved from the pristine single crystal. **b** Magnified image of the active part of the device. Two sets of nozzles (on the left and lower edge) emit the electrons either along the flat side or corner of the hexagonal Fermi surface (overlaid). This central region of the crystal was thinned down from 10 μm to 1.2 μm. **c** Magnification of the lower nozzle set. Each of the nozzles is approximately 250 nm wide and they are spaced about 1 μm apart. **d** Schematic diagram of the measurement setup. A current is applied between the top left electrode and one of the nozzles while the voltage is measured between the top-right electrode and another nozzle. **e** Non-local resistance measurement as a function of temperature of the setup in panel d. The obtained voltage values are in good agreement with the results of finite element simulations which assume bulk resistivity values and solve the Laplace equation in this geometry

crystal (darker grey square, magnified in Fig. 3.28b) was thinned with the FIB from 20 to about 1.2 μm, which is necessary to achieve the desired aspect ratios when milling the closely spaced constrictions in Fig. 3.28c. Each of the nozzles is roughly 250 nm wide and the smallest spacing between them is 1 μm. A schematic diagram of the device is presented in Fig. 3.28d. There are two sets of nozzles (1–8) and (A–D) which are arranged perpendicularly to each other. In this way TEF is probed with different orientations relative to the underlying Fermi surface, namely in the 2-beam and 3-beam directions. The sample boundaries in between the nozzles and along the central part of the TEF devices were polished in the FIB with a low current to minimise surface roughness. The two large electrodes are placed far away from the nozzles so that they do not influence the TEF signal even when the cyclotron radius is large (at low fields). The temperature dependent non-local voltage signals measured between V_{com} and the contacts B, C and D in response to a current sourced from

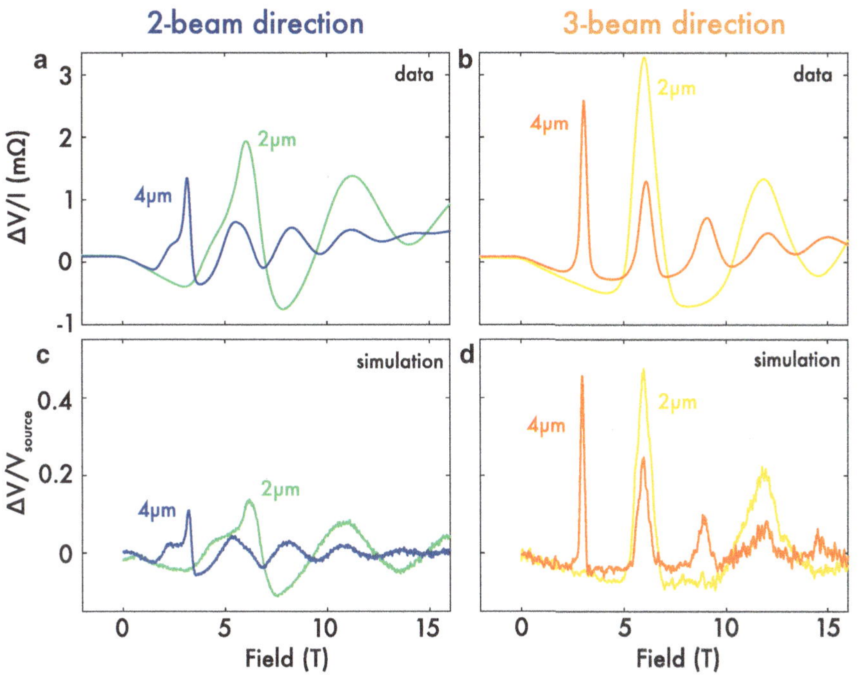

Fig. 3.29 Comparison of TEF measurements and simulations. **a** and **b** are a direct comparison of the TEF spectra obtained from nozzle pairs which are separated by 2 μm or 4 μm. While the 3-beam direction displays pronounced single peaks, the amplitude of the 2-beam direction is lower and the peak is followed by a broad hump. **c** and **d** are Monte Carlo based simulation results which are in excellent agreement with the data. The Monte Carlo simulations were performed by Aaron Sharpe

contact A, are displayed in panel Fig. 3.28e. Because the device size is larger than the mean free path at all temperatures, the non-local voltages can be compared to the results of finite element analysis, which calculate the solution to the Laplace equation for this particular device geometry based on bulk resistivity measurements [16]. The results are in good agreement with the measurements, indicating the high-quality of the single crystal.

The transverse electron focusing spectra measured at 2 K along the 2-beam and 3-beam directions are presented in Fig. 3.29a and b. In the 2-beam configuration the current was sourced from contact 1 and drained to the top left electrode, while for the 3-beam orientation contact A was used as the source. The voltages between V_{com} and the remaining contacts were measured as a function of magnetic field. For clarity only the voltages measured at distances of 2 μm and 4 μm are displayed. The 2-beam direction shows a distinctive side peak feature, while the 3-beam direction is characterised by a pronounced single peak. The background signal, which is similar along both directions, is negative over most of the field range. This voltage inversion is explained by the width difference of the voltage contacts. Because the width of the nozzles is only 250 nm, electron skipping orbits can easily 'miss' a contact if the

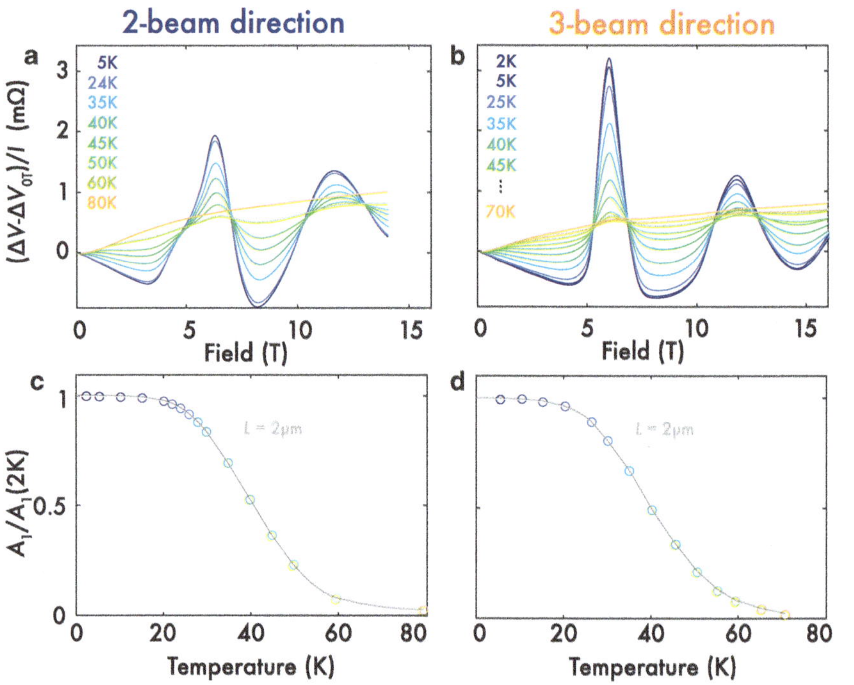

Fig. 3.30 Temperature dependence of the 2-beam (**a**) and 3-beam (**b**) TEF spectra for nozzle pairs spaced by 2 μm in both cases. The temperature dependent zero-field voltage was subtracted from all curves. Panels **c** and **d** show the normalised amplitudes of the first peak. The grey line is a guide to the eye. Both directions show a nearly temperature independent between behaviour 2 K and 20 K and then gradually decrease between 20 K and 70 K; a temperature dependence which is reminiscent of that of the mean free path in Fig. 3.7

cyclotron diameter is much larger than the nozzle width. The voltage electrode V_{com} on the other hand is much larger, such that all the electrons are counted, which is the reason for the negative voltage background. Further, Monte Carlo simulations of 30000 ballistic electrons propagating in $PdCoO_2$ were performed by Aaron Sharpe and are displayed in Fig. 3.29c and d. His code is based on the experimentally determined Fermi surface [29] and neglects bulk scattering. The agreement between the simulations and the data is remarkable. The peak shapes and background along both directions are reproduced including many details. The biggest discrepancy between the experimental data and the simulation arguably lies in the difference of the relative signal sizes between the two orientations. For instance, for a nozzle separation of 4 μm, the simulations predict a difference in peak height of around a factor 3, while the experiment finds a difference of approximately 1.6. This may be a consequence of the assumption that the electrons are ejected from the nozzle with a $\cos(\theta)$ distribution, which may not necessarily be the case, in particular along the 2-beam direction.

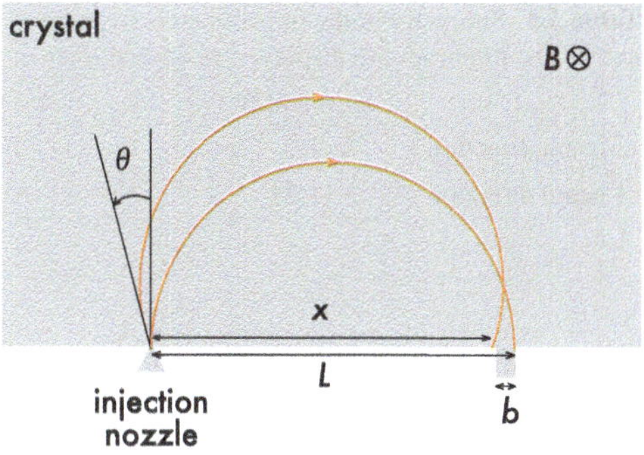

Fig. 3.31 Geometrical model of TEF on a circular Fermi surface. The grey shaded region represent the single crystal device which is placed into a perpendicular magnetic field B. In a simplified picture, the injection nozzle is assumed to be a point like emitter, while the probing nozzle has a finite width b

The temperature dependence of the mean free path l in $PdCoO_2$ was shown in Fig. 3.6. Around 50 K the mean free path suddenly increases and becomes as large as 20 μm below 20 K. A similar behaviour is observed for the temperature dependence of the TEF peak amplitude displayed in Fig. 3.30. Panel a and b show the TEF signal in the 2-beam and 3-beam directions from pairs of nozzles which were separated by 2 μm. The peak height from the main focusing peak A_1 normalised at 2 K is plotted as a function of temperature in panels c and d. Both directions have the same temperature dependence, reminiscent of the temperature dependence of the mean free path. This indicates that the observed TEF signal is directly connected to the mean free path of the electrons, as expected. Lastly we remark that the TEF signal is not sensitive to the electron phase and so no statement about the phase coherent mean free path can be made based on these measurements.

3.4.4 Extracting the Mean Free Path

Rather than rely purely on analysis of the zero field resistivity, the mean free path can also be deduced, with only a few assumptions, from the TEF data. The first peak in the TEF spectrum is created from electrons which are directly focused from the current injection nozzle in to the voltage probing nozzle without scattering from the boundary. When the nozzle distance L is increased, the path length s through the samples increases accordingly, such that the effect of bulk scattering on the TEF amplitude can be probed. With this analysis the mean free path l can be extracted based on geometric arguments and quantities alone. Tsoi et al. [86] noted that the amplitude A_1 of the first TEF peak is proportional to $e^{s/l}$ multiplied with a function $f(b/L)$, which depends on the ratio of the nozzle width b to the nozzle separation L. The function f accounts for the fact that, for equally wide nozzles, fewer electrons hit a nozzle the further away it is from the injection point. In other words, the maximal angle θ at which an electron can be ejected, such that it will still hit the distant nozzle decreases the larger the separation L becomes. The functional form of $f(b/L)$ for a circular Fermi surface can be found quickly and is derived in the next paragraph.

Table 3.3 Summary of the fit parameters from fitting equation (3.14) to the TEF data in Fig. 3.32a and b. The fit curves are displayed as dashed lines in panels c and d of Fig. 3.32

	A (mΩ)	t (–)	l (μm)
2-beam direction	7.7	−0.0037	14
3-beam direction	11.24	0.06	15.7

Assuming a point-like injection source and a voltage probing nozzle with width b, c.f. Fig. 3.31, we find the minimal focusing distance $x = b + L$, where L is the maximum distance an electron can travel at a fixed field ($L = 2r_c$). The angle θ_{max}, which focuses an electron on to the point x can be determined by trigonometry and is $x = 2r_c \cos(\theta_{max})$. By Taylor expanding $\cos(\theta) \approx 1 - \frac{\theta^2}{2}$ for small θ, and combining the above equations we find $b + L \approx L\,(1 - \frac{\theta_{max}^2}{2})$ and hence $\Delta\theta = 2\theta_{max} \approx 2\sqrt{2b/L}$. With this we conclude that in a system with a circular Fermi surface the amplitude of the first peak scales with nozzle spacing as $A_1 \propto e^{s/l}\sqrt{2b/L}$, where the path length s is given by half the circle circumference $s = \pi r_c$. Note that this analysis is only valid for small values of $\frac{b}{L}$, i.e. $b \ll L$ such that the Taylor expansion around small values of θ is a good approximation. If the nozzle width b becomes comparable to the nozzle separation L, the above analysis breaks down and should no longer be expected to describe the amplitude decay dependence.

In the case of a hexagonal Fermi surface the amplitude A_1 depends similarly on $e^{s/l}$, however the functional form $f(b/L)$ will be modified. In order to fit the peak decay and extract the mean free path l along the 2-beam and 3-beam directions in $PdCoO_2$ from the results in Fig. 3.32 I will use the form

$$A_1(L) = Ae^{s/l}\sqrt{\frac{2b}{L} + t}, \tag{3.14}$$

with the nozzle width b=0.3 μm and $s_{2beam} = \frac{3}{2}L$, $s_{3beam} = \sqrt{3}L$ are the path lengths for the 2-beam and 3-beam directions respectively. Apart from l, the free variables are A, which sets the overall amplitude and t, which takes into account the Fermi surface shape deviations compared to the circular case. In the 3-beam direction the path length is ill-defined due to the very nature of the super-geometric focusing effect. We choose the average between the longest and shortest possible paths. The results of fitting equation (3.14) to the experimental TEF data in Fig. 3.32a and b are summarised in Table 3.3 and plotted in Fig. 3.32c and d.

From the fit parameters in Table 3.3, we can conclude that

- the super-geometric focusing effect does indeed increase the overall amplitude A of the focusing peaks along the 3-beam direction compared to the 2-beam direction.
- along the 2-beam direction the focusing amplitude decays similarly as for a circular Fermi surface, as evidenced by the small, nearly zero value of t. This is not surprising, as the rounded corners of the hexagon can locally be approximated by

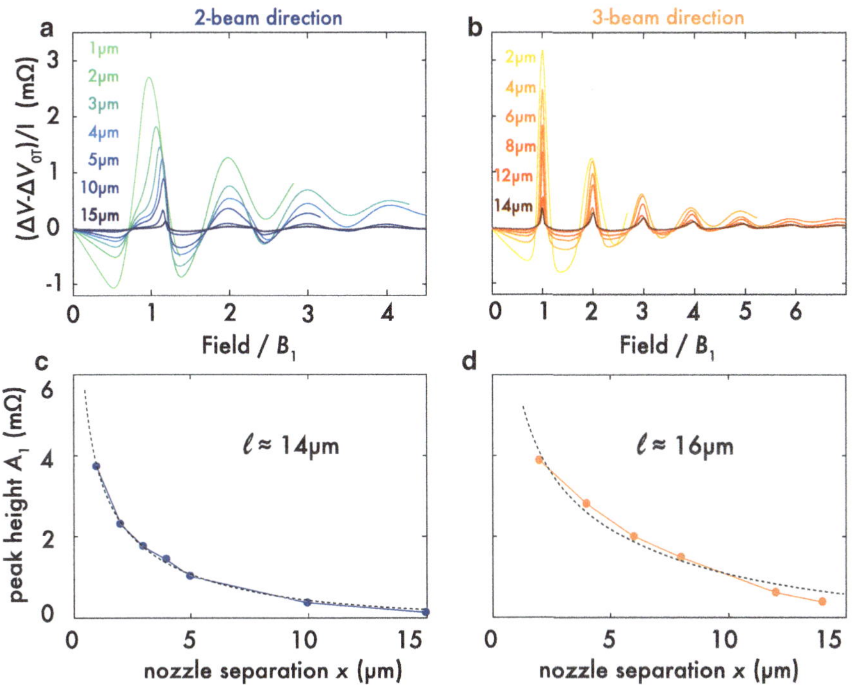

Fig. 3.32 Mean free path extraction from the amplitude decay of the TEF peaks as a function of nozzle spacing. **a** and **b** show the TEF spectra of differently spaced nozzle pairs scaled according to the position of their peaks. With increasing nozzle separation, the path length of an electron trajectory through the sample grows and the chance of an electrons being scattered away from their focusing orbit is enhanced exponentially. By fitting the decay of the primary peak amplitude A_1, a mean free path of $l \sim 15\,\mu$m can be extracted **c** and **d**. The dashed line is a fit to the data based on Eq. (3.14). The fit parameters are summarised in Table 3.3

a circle. Hence the transverse focusing behaviour along the 2-beam direction is in essence very similar to the situation in a circular Fermi surface system.

- in the 3-beam direction, the additional non-zero t term leads to a slower decay of the peak amplitude compared to a system with a circular Fermi surface with the same mean free path l (c.f. dashed curve in Fig. 3.32d). This implies that the collimation of the ballistic electrons caused by the flat Fermi surface facets leads to a boost of the ballistic effects, such that TEF can be detected over a longer length scale along the 3-beam direction than in comparable devices with a circular Fermi surface.
- the mean free path $l \approx 15\,\mu$m is essentially the same along the 2-beam and 3-beam directions and is in good agreement with the value extracted from the in-plane resistivity measurement discussed in Fig. 3.7.

By comparing the TEF spectra measured between nozzles of varying separation L in Fig. 3.32a and b, we discover that the peak shapes are self-similar for the most

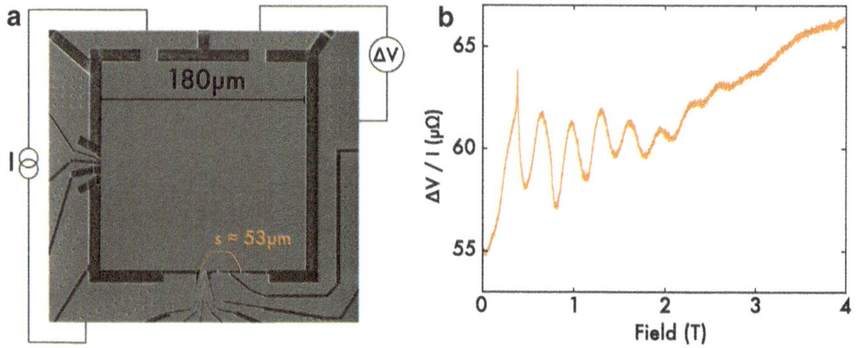

Fig. 3.33 Maximum measured TEF spectrum. **a** The nozzles are spaced by $L = 35\,\mu m$, yielding a path length of $s = 53\,\mu m$ through the device along the 2-beam orientation. **b** Up to 8 focusing maxima are detectable when measured at $T = 1.8\,K$

part. However as the separation L becomes smaller and approaches the size of the nozzle width b the features in the peaks, especially along the 2-beam direction are broadened. This is a result of the fact that the maximum ejection angle that projects an electron onto the distant nozzle $\Delta\theta$ decreases as $\propto \sqrt{\frac{b}{L}}$. Or put differently, if the nozzle width b were scaled with the nozzle separation L the TEF peaks would recover self similarity.

In panel c and d of Fig. 3.32 the amplitude of the primary focusing peak A_1 is plotted as a function of the respective nozzle separation. The dashed line is a fit with the functional form described in Eq. (3.14) with the fit parameters in Table 3.3. The fit works well for the data measured along the 2-beam direction but shows some deviations from the data measured in the 3-beam orientation. This might be due to an invalid assumption about the ill-defined path length s_{3beam} or the functional form of $f(b/L)$ in the super-geometric orientation.

3.4.5 Maximal Transverse Electron Focusing

As the name suggests, the mean free path l of a system describes the average distance an electron can propagate between scattering events which modify its direction. Since this is a stochastic process, a fraction of the electrons can propagate undisturbed for longer distance and lead to a TEF signal although the path length s through the sample exceeds the mean-free path. Here I present measurements of the TEF signal where the path length s is 3.5 times longer than the mean free path l. The corresponding device and measurements are shown in Fig. 3.33. The focusing device is the same as the one introduced in Fig. 3.28. The voltage probing nozzle is $L = 35\,\mu m$ away from the injection nozzle, defining a path length of $s = 53\,\mu m$ in the 2-beam orientation. Although the signal size is reduced by two orders of magnitude compared to the closely spaced nozzles in Fig. 3.29, the voltage peaks are clearly resolvable and 8 focusing peaks are detected. For the closer spaced nozzles the width to separation ratio was about $b/L \sim 250\,\text{nm}/2\,\mu\text{m} \sim 12\%$, while here it is less than 1%. This decreases the acceptance angle $\Delta\theta \propto \sqrt{\frac{b}{L}}$ of electrons which are directly focused from one nozzle into the next. As a result the TEF spectrum shows a pronounced and sharp primary TEF peak. As the TEF signal is still several times larger than the noise level it would be feasible to further increase the nozzle spacing. Moreover it will be informative to study this effect in the super-geometric focusing orientation, where we expect an even stronger response at these length scales.

3.4.6 Influence of the Nozzle Width on the Focusing Peak Amplitude

To systematically investigate the dependence of the TEF signal amplitude on the nozzle width, four sets of nozzles with constant separation $L = 3.5\,\mu\text{m}$ and nozzle widths $b = 270\,\text{nm}$, 430 nm, 500 nm, and 700 nm were fabricated side by side into the same crystal device (Fig. 3.34a and b). The TEF spectra, measured at 2 K are presented in Fig. 3.34c.

As the nozzle width is reduced the amplitude of the TEF peaks grows and at the same time their full width at half maximum (FWHM) decreases, as expected. If electron focusing is measured between nozzles of different widths, their average value approximately determines the height and FWHM of the focusing peak as evidenced by the grey trace in panel (c). There TEF was measured between nozzles that are 270 nm and 430 nm wide and the corresponding peak amplitude is consistent with a value for a $\approx$350 nm wide nozzle pair based on the fit to the amplitude decay in panel (d). Extracting the peak heights and plotting them as a function of their average nozzle width b in panel (d), we find that amplitude is approximately inversely proportional to the nozzle width, as indicated by the dashed curve, which shows a power law fit. Upon close inspection of the data in panel (c) we find that the top electrodes are spaced closely enough to yield a focusing peak around -1 T, even though the nozzles are each approximately $5\,\mu\text{m}$ wide. The electron orbits are indicated in panel (a). Because the top electrodes are separated by $\approx 10\,\mu\text{m}$ rather than $\approx 3.5\,\mu\text{m}$, as is the case for the other nozzle pairs, we expect the peak amplitude to be suppressed by roughly a factor 3 based on the analysis in Fig. 3.32d and correct for this in panel (d). The slow decay of the amplitude for large values of b, which may be surprising at first, is likely to be a consequence of the super-geometric focusing effect.

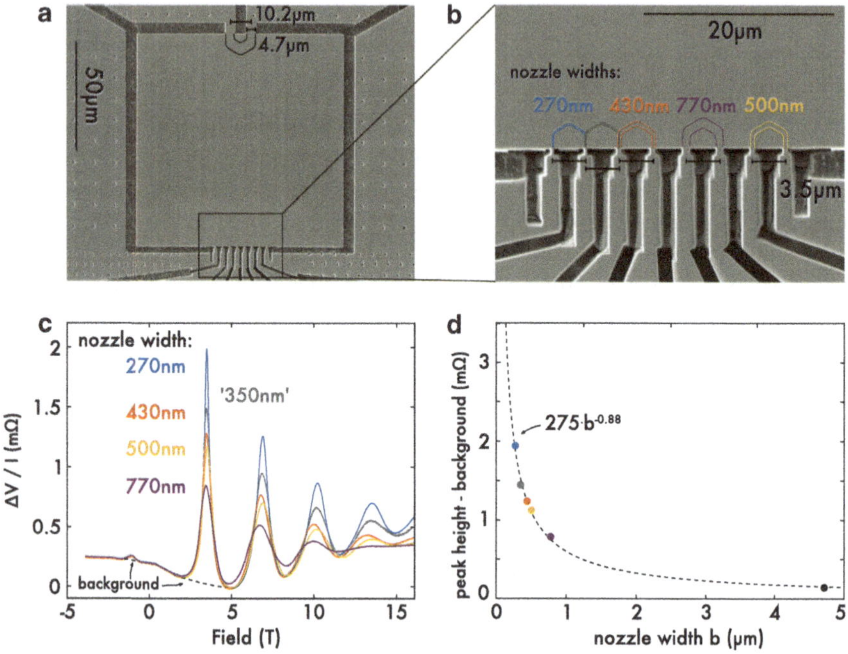

Fig. 3.34 **a** Overview of the TEF device, which consists of four pairs of nozzles each separated by 3.5 μm (centre to centre) oriented along the 3-beam direction. **b** The nozzles widths vary from 270 nm to 770 nm. The coloured traces represent the smallest and largest cyclotron orbits which fulfil the focusing condition. **c** Comparison of the TEF spectra measured at 2 K. The grey curve was measured between two nozzles of different widths (270 nm and 430 nm) and hence 350 nm denotes their average. The top electrodes are spaced closely enough that focusing can be detected at −1 T. **d** The peak amplitudes (measured from the background indicated by the dashed curve) of the first focusing peak are plotted as a function of the nozzle width b (solid circles). The TEF amplitude of the top electrodes (black circle) was multiplied by a factor of 3 to correct for the larger nozzle distance between the top electrodes (10.2 μm) compared to the nozzles (3.5 μm) based on the analysis in Fig. 3.32d. A power law decay of the form $A \cdot b^c$ can be used to describe the decay of the TEF peak height as a function of increasing nozzle width b. The black dot at $b = 4.7\,\mu$m was not included in the fit

3.4.7 Specularity of the Boundaries

In the purely ballistic regime, where the only dominant scattering events are the interactions of the electrons with the sample edge, the macroscopic transport properties depend on the microscopic details of the boundaries. For a microscopic description, the electron Fermi wave length λ_F and the surface roughness of the boundary are the relevant length scales in the problem. An alternative, macroscopic description of the boundary is attained by defining a parameter p, which describes the probability of an electron wave packet being reflected specularly from the boundary Consequently the probability of diffusive scattering is then given by $(1 - p)$. Based on this phenomenological parameter p many theories (c.f. Sect. 3.2.3) have been developed which describe the electronic transport in the ballistic regime. Moreover the specularity coefficient p is often used to describe the quality of a surface. A popular method of extracting the surface parameter p is by calculating the ratio of subsequent TEF amplitudes $q = \frac{A_{n+1}}{A_n}$ and equating p with q. In the following I will present simulations which show that equating p with q is an indication, but by no means an exact quantification of the specular properties of the boundary, in particular along the super-geometrical focusing orientation. Further, with the results of the simulations I will attempt to quantify the specularity of the FIB fabricated boundaries of the $PdCoO_2$ devices presented here.

It is straightforward to extend the simulations of the TEF spectra presented in Fig. 3.26 to include a varying boundary specularity coefficient p between 0 and 1. One example of such a spectrum, which was simulated for 100000 particles along the 3-beam direction for $p = 0.9$ is displayed in Fig. 3.35a. As the distance x from the injection nozzle increases, the peak amplitude decreases. Upon close inspection of the spectrum a double peak substructure is observable. This is an artefact of the Fermi surface parametrisation based on a harmonic expansion, which led to a slightly concave hexagonal facet. Neither the ARPES [25] nor this TEF experiment have the necessary resolution to resolve the warping, so I will treat it as a artefact of the simulation. Because of the divergence of the focusing peak it is numerically difficult to define an absolute peak height, so instead the total number N_n of electrons which form the nth focusing peak are counted (orange shaded regions in Fig. 3.35a). The ratio between two subsequent peaks N_n/N_{n+1} as a function of peak number is displayed in panel b. From this graph we can make two deductions:

- as evidenced by the peak number independence of the peak ratio q (i.e. because the curves are constant), the amplitude of TEF peaks follow a power law decay:

$$\frac{N_{n+1}}{N_n} = q \quad \Rightarrow \quad N_n = q \cdot N_{n-1} = q \cdot q \cdot N_{n-2} = q^n \cdot N_0.$$

- the peak ratio q is not equal to boundary specularity p, especially for low values of p.

Having established that the peak ratios follow a power law decay, this can be used for fitting the peak decay and extracting the q-value. A function of the form $N_0 \cdot q^n + c$ was used to extract the q-values for the 3-beam and 2-beam orientation as well as for a circular Fermi surface for values of p between 0 and 1. In this fit, N_0 scales the overall ratio, q is the extracted amplitude ratio and c accounts for the background offset. The results are presented in Fig. 3.35c. The main conclusion is that the simple assumption of $q = p$ is not strictly applicable, even for a circular Fermi surface. The reason for this is two-fold. Firstly, the TEF peak becomes wider as n increases, such that the measured voltage decreases if the nozzle width b is constant even in the case of perfectly specular boundaries. Secondly, even for diffusive boundaries a macroscopic number of electrons will be statistically back-reflected into an angle $\Delta\theta$ leading to a further focusing peak. This effect is particularly pronounced in the super-geometrical

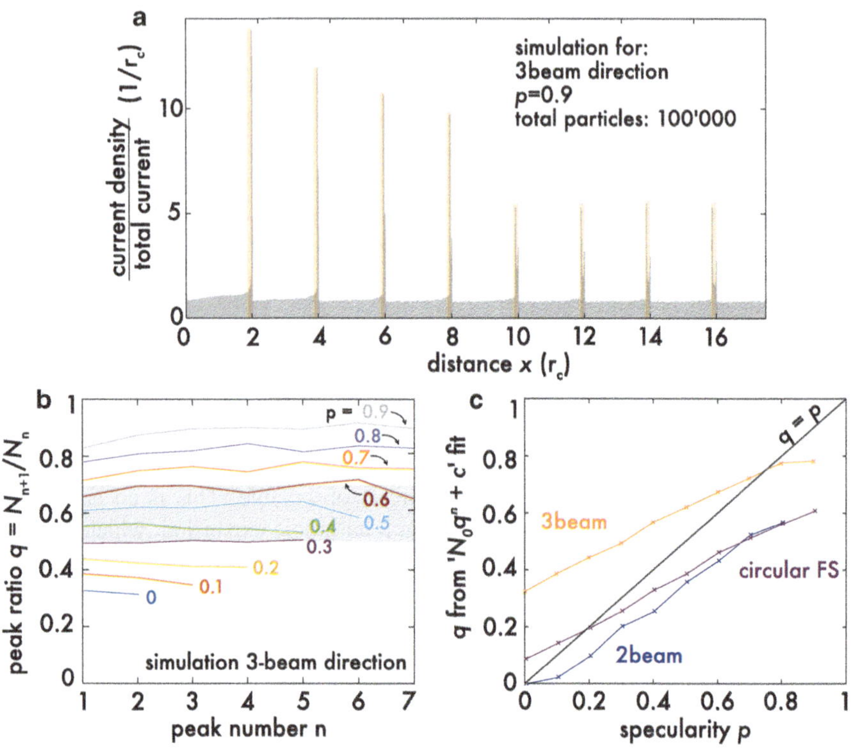

Fig. 3.35 **a** Simulated TEF spectrum for 100000 particles which were injected along the 3-beam direction. The surface specularity coefficient was set to $p = 0.9$, indicating that 90% of the electrons were specularly reflected. **b** The extracted peak ratios q as a function of peak number n. **c** The extracted q-values from a power-law fit of the TEF peak decay for the 3-beam and 2-beam orientation as well as for a circular Fermi surface

focusing orientation, which leads to a strong increase of the q values. Consider for instance the 3-beam direction in the case of completely diffusive boundaries ($p = 0$). The simulations in Fig. 3.35c yield a q-value of approximately 0.33. This is due to the restriction of the ballistic motion on a hexagonal Fermi surface, which causes approximately 1/3 of the electrons to propagate in the direction they would have if they were specularly reflected.

Based on the results of the simulations above and the TEF measurements presented in Fig. 3.32, in this section I will attempt to quantify the specularity of the FIBed boundaries in $PdCoO_2$. A typical TEF spectrum is displayed in Fig. 3.36a. In order to extract the peak heights A_n, a background, indicated by the dashed line, is defined by connecting neighbouring voltage minima. The resulting ratios of adjacent peak amplitudes along the 2-beam and 3-beam directions are presented in panels (c) and (d) respectively. From this the average q-value can be determined to be $q = 0.6 \pm 0.1$ (grey shaded region). Comparing this to the results of the simulations in Fig. 3.35c,

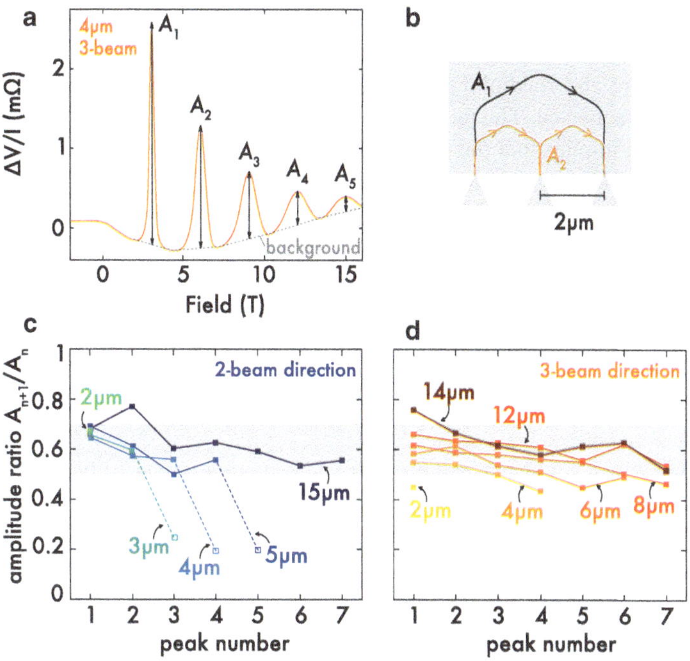

Fig. 3.36 **a** A typical TEF spectrum measured along the 3-beam direction for a nozzle pair with 4 μm nozzle separation. **b** Sketch of a set of TEF nozzles. At the position where the electrons forming the second TEF peak impact the boundary there is another (floating) nozzle. This does not change the TEF spectrum. **c** and **d** show the experimentally extracted amplitude ratios $q = A_{n+1}/A_n$

we find the boundary specularity coefficient p of FIB defined $PdCoO_2$ edges is between 0.3 and 0.6.

Further we notice the open box symbols in panel Fig. 3.36c which clearly lie below the average amplitude ratio values. This occurs when the TEF peak amplitude was determined in a field region where the cyclotron radius r_c is on the order of the nozzle width b. In $PdCoO_2$ this occurs roughly above $B = \frac{\hbar k_F}{er_c} = \frac{\hbar\, 0.96\,\text{Å}^{-1}}{e\, 500\,\text{nm}} \approx 12.6\,\text{T}$, i.e. when the cyclotron radius $r_c = 500$ nm. Remarkably the focusing peaks along the super-geometric focusing direction do not seem to suffer from the same effect. Naturally once the cyclotron radius is smaller than the nozzle width the electron skipping orbits will hit both voltage contacts and a TEF peak will no longer be observable.

I will conclude the section on the specularity of the boundaries by discussing another feature of the focusing peaks, or rather lack thereof. As sketched in Fig. 3.36b the TEF devices consist of a set of equidistant nozzles. This can lead to the situation that skipping orbits, which lead to higher order focusing events, impact the boundary right at the position of another nozzle. This is the case, for instance, if transverse focusing is measured between nozzles separated by 4 μm in a device which has series of 2 μm spaced nozzles. Naively one might expect the second A_2 and fourth peak

A_4 to show a reduced amplitude, which is not the case as evidenced by the data in Fig. 3.36a. The centre contact is electrically floating and hence an equal number of electrons are absorbed and emitted by the charged centre nozzle once a steady state has been achieved. The fact that the presence of the nozzle does not modify the TEF spectrum implies that the emission spectrum of the nozzle is equivalent to the spectrum of back-reflected electrons by the sample boundary, an observation which possibly calls into question the entire interpretational structure of TEF as a probe of surface roughness.

3.4.8 Real Space Propagation of Electrons in the Ballistic Regime

In this section my goal is to explore the real-space trajectories of ballistic electrons subject to a magnetic field in a TEF device. How valid is the semi-classical picture of electrons skipping around the sample edge on hexagonal orbits? The devices are in the quasi-ballistic regime, meaning that the distance between the nozzles is less than the mean free path l, such that ballistic effects become observable in a magnetic field; however the distance between both voltage contacts is greater than l, such that a diffusive contribution to the voltage signal is expected as well.

Taking a step back, let us consider the situation in the ohmic regime of a simple metal, for instance copper. Imagine a transport bar in which the current is flowing parallel to the axis of the bar, in a perpendicular magnetic field. The Lorentz force causes the electrons in the system to propagate to one side of the sample causing a potential difference, the Hall voltage, between opposite sample edges. The voltage creates an internal electric field which opposes the Lorentz force. As a consequence, electrons experience no net force when passing through the sample in the steady state.

So are the electrons in a TEF experiment actually propagating around the device edge from nozzle to nozzle, or are we measuring the complicated voltage distribution created inside the sample in order to create a no net force situation for the electrons? In fact, the description from above breaks down in a ballistic system where the movement of electrons can no longer be described by gradients in the electric field. When the elastic mean free path is much larger than the sample, then a Landauer–Büttiker description becomes appropriate, which describes the transport of charge carriers in an electric field free ($\vec{E} = 0$) sample in terms of skipping orbits along the sample edge.

To address this question experimentally, I have fabricated a TEF device with a hole which is designed to obstruct the path of the electrons between to focusing nozzles. The device is presented in Fig. 3.37a. The electron trajectories outlined in yellow are the ones which lead to the first focusing peak. The position of the hole is such that the trajectories which lead to higher order focusing peaks can pass underneath. The TEF spectrum of the device with and without the hole are shown in Fig. 3.37b.

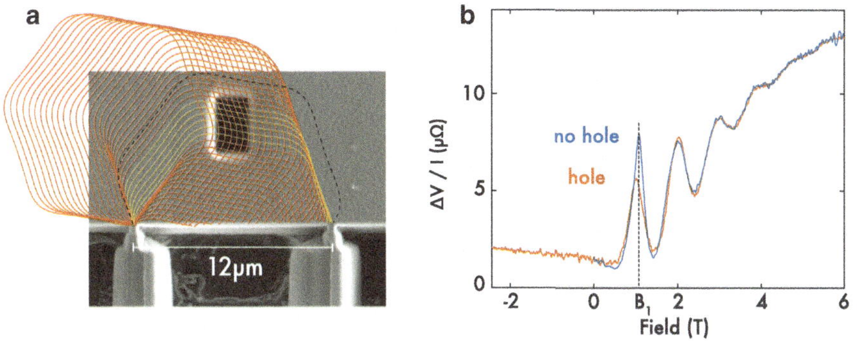

Fig. 3.37 **a** A hole was milled into a TEF device after it was characterised in absence of the hole. The hole is positioned such, that the ballistic electron paths which lead to the first focusing peak (marked in yellow) are disrupted, while the trajectories for higher order focusing events with smaller cyclotron radii can pass underneath. The device was cut 9° off axis and hence the cyclotron trajectories are slightly bent. **b** The TEF spectra with and without hole in comparison. While the amplitude of the first peak is clearly reduced, the higher order focusing events are completely unaffected by the presence of the hole

Clearly the first focusing peak amplitude has decreased, while the higher order peaks remain unaffected. The reason why first the peak has not completely vanished is that the hole was not cut big enough. At fields below B_1 electron orbits such as the one indicated by dashed black line in panel a still hit the second focusing nozzle. To test this hypothesis, the straightforward experiment will be to enlarge the extent of the hole and repeat the measurement. Strictly speaking, this experiment in which a voltage was measured under constant current bias of the sample, does not prove the point of circulating edge currents, because a measured potential difference does not directly yield information about the spatial flow of current. The direct way of demonstrating circulating edge currents would be to measure the current flow under constant voltage bias of the sample.

A similar focusing device which challenges our understanding of the electron-boundary interaction is displayed in Fig. 3.38a. Instead of a polished boundary between the nozzles, there is a trough which is electrically floating. The idea of this setup is to probe the influence of the trough on the second focusing peak. Naively one might expect the second peak to vanish completely, as there is no boundary to back-reflect from. Remarkably however, the second peak is still clearly detectable as shown in panel b. Perhaps even more surprising is the fact that the amplitude ratio A_2/A_1 is identical with and in the absence of a trough between the nozzles as evidenced by the dashed red trace. This result is seemingly at odds with the common interpretation of the amplitude ratio as the boundary specularity coefficient p, c.f. discussion about Fig. 3.35. Further experiments with troughs between nozzles are definitely required to gain further insight into this phenomenon.

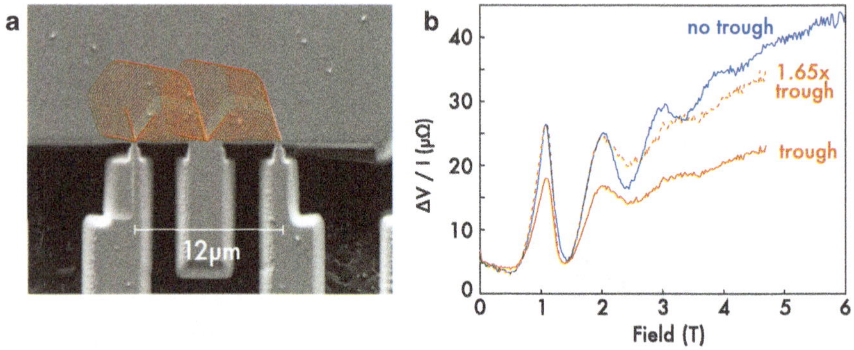

Fig. 3.38 **a** Instead of a polished device boundary, a trough was left in between two focusing nozzles. The device was cut 9° off axis. The yellow traces represent the electron orbits which should back-reflect at the missing device boundary and form the second focusing peak. **b** Comparison of a conventional TEF signal to TEF spectrum where a trough was carved in between the focusing nozzles. For a direct comparison of the amplitudes the TEF spectrum from the device containing a trough was scaled by a factor of 1.65 (red dashed curve)

3.4.9 A Landauer–Büttiker Description of TEF

In Sect. 3.2.5 we saw that a Landauer description of the ballistic transport becomes appropriate if the elastic mean free path of the electrons in a system is larger than the system size. The central idea of this formalism is to describe the transport of electrons through a device by dissipationless edge channels which connect adjacent ohmic contacts. In the *linear response* regime a four-terminal system with current contacts i and j and voltage probes k and l is characterised by

$$\begin{pmatrix} I_i \\ I_j \\ I_k = 0 \\ I_l = 0 \end{pmatrix} = G \cdot \begin{pmatrix} V_i \\ V_j \\ V_k \\ V_l \end{pmatrix}, \tag{3.15}$$

where the conductance G is a 4x4 matrix which is independent of the contacts i, j, k, l. Based on time-reversal invariance and current conservation, Büttiker [67] derived the reciprocal relation for the conductance

$$G_{ij,kl}(B) = G_{kl,ij}(-B), \tag{3.16}$$

which is known as the generalised Onsager relation. In other words, the conductance of a system is invariant under reversal of the magnetic field direction if the current and voltage leads are exchanged simultaneously. Another property of a four-terminal device in the linear regime is that the voltage response is odd under current reversal

$$V_{kl}(I_{ij}) = -V_{kl}(-I_{ij}). \tag{3.17}$$

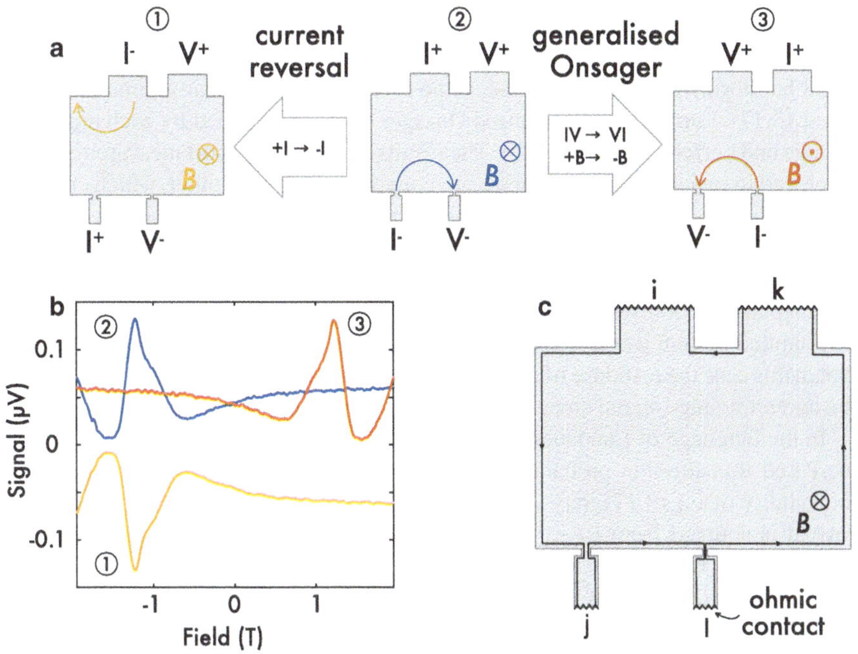

Fig. 3.39 **a** In a dc transport experiment the conventional TEF setup ② is modified by ① reversing the direction of the current ($+I \rightarrow -I$) and by ③ exchanging the current and voltage contacts ($+I \rightarrow +V$; $-I \rightarrow -V$). **b** The resulting TEF spectra of the 3 variations. **c** A Landauer–Büttiker picture of the electronic conduction in the quantum Hall limit. One-dimensional edge modes connect adjacent ohmic contacts (zig-zag lines)

This is a direct consequence of the linearity of Eq. (3.15), which is directly shown by multiplying both sides with $\begin{pmatrix} -1 & 0 \\ 0 & -1 \end{pmatrix}$. A convenient consequence of the fact that the voltage response is odd under current reversal, is that ballistic transport measurements in the linear regime can be performed with a standard ac lock-in technique. I emphasise that the relations (3.16) and (3.17) only strictly hold in the limit of linear response. Non-linearities which violate (3.17) may occur due to quantum interference effects or inelastic scattering processes [101–103]. Therefore, when measuring with an ac transport setup it is important to check that the voltage response does not contain higher-harmonic contributions generated by non-linear processes. In particular the loss of current reversal symmetry will generate second harmonics and hence a convenient way of confirming the linear regime is by simultaneously measuring the second-harmonic response of the system during a transport measurement.

In $PdCoO_2$ inelastic electron-phonon or electron-electron collisions are rare at low temperatures such that the requirement of a long elastic mean free path $> L$ is fulfilled. First, I will confirm the validity of the Eqs. (3.16) and (3.17) by presenting the results of a dc TEF transport experiment. Thereafter I will discuss transverse electron focusing in a Landauer–Büttiker framework.

The standard TEF setup ② alongside two modified situations are sketched in Fig. 3.39a. By reversing the polarity of the current (② → ①), we expect to find an equal but negative voltage response, if the system is in the linear regime according to Eq. (3.17). Further, the generalised Onsager relation is probed by exchanging the voltage and current leads (② → ③). The results of the dc transport measurements are displayed in panel b. All three curves are congruent with each other, which confirms that the system is in the linear response regime and can be measured with an ac technique. The dc measurements were performed with a Keithley 6221 current source and a 2182 A nanovolt meter. The delta option, which sources current in form of a rectangular function was used to avoid undesired voltages from thermal gradients. The applied current was switched between 0 and $+I$ or 0 and $-I$ respectively. Note that in this case the response of the Keithley must be multiplied by a factor 2 to obtain the correct voltage signal size.

In the language of Landauer–Büttiker, transverse electron focusing describes an increased transmission probability between two adjacent contacts. At first glance the validity of Eq. (3.17) may seem counter intuitive. Why does it not matter if the current is sourced from $i \to j$ or from $j \to i$ if the voltage contacts are only on one side? The reason is that the unidirectional channels are dissipationless and circulate the *entire* edge of the sample, as sketched in Fig. 3.39c. The modes only depend on the magnetic field strength and polarity, as well as the geometry of the device. Hence the situations sketched in ① and ② of Fig. 3.39a actually belong to the *same* Landauer–Büttiker picture, with reversed electrochemical potentials in the leads. Therefore, upon current reversal the voltage response is identical up to a sign change. In contrast, reversing the magnetic field direction changes the quantisation condition which determines the shape of the electron trajectories. Because the contact leads and hence the transmission probabilities of the contacts are not symmetric, the TEF signal is only symmetric under field reversal if the current and voltage contacts are exchanged, just as the generalised relations dictate. Lastly I remark that unlike in semiconducting 2DEGs where the number of edge modes can be reduced to only few, typically of order 1–10, the number of edge modes in a metal such as $PdCoO_2$ contributing to the TEF signal is on the order of 10^3 per palladium layer. This was estimated through the first focusing peak A_1 displayed in Fig. 3.36a of a device with thickness $t = 1.2\,\mu\text{m}$. The amplitude of A_1 is $\Delta R = 2.8\,\text{m}\Omega$ and the sample has $n = \frac{1.2 = \mu m}{17.73\,\text{Å}/3} = 2032$ conducting palladium layers in parallel, such that the resistance contribution per Pd layer is $R_{\text{Pd}} = n \cdot \Delta R = 5.69\,\Omega$. The associated conductance $G_{\text{Pd}} = R_{\text{Pd}}^{-1} \approx 2300\ G_0$ reveals the number of modes, which each contribute a conductance quantum G_0 to the focusing peak.

3.4.10 The Effect of Small Device Dimensions

So far we have investigated TEF devices in which the distance between the nozzles and the top electrodes was much larger than the mean free path l. The voltage differ-

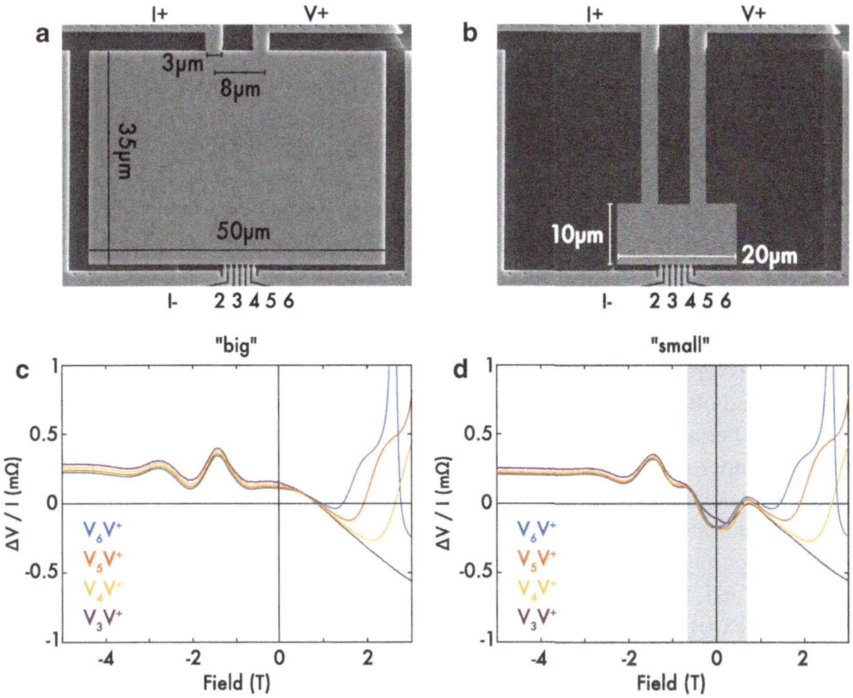

Fig. 3.40 Size dependent voltage inversion in zero-field. **a** and **b** show the same focusing device before and after its dimensions were reduced from $35\times50\,\mu\text{m}^2$ ("big") to $10\times20\,\mu\text{m}^2$ ("small"). **c** and **d** display the corresponding TEF measurements. The conventional focusing peaks between the nozzles are detected at positive fields and focusing between the top electrodes lead to peaks at negative fields. The "big" device shows no anomaly in zero-field, while the "small" device shows a pronounced negative peak

ence in the absence of a magnetic field ($r_c = \infty$) is determined by the solution of the Laplace equation. By reducing the spacing between the current injection contact (I^-) and the voltage electrode (V^+) to below the mean free path l a diffusive description of the situation is no longer valid and ballistic effects can influence the measured voltage difference.

In the following I will present an experiment where the FIB was used to reduce the area of a TEF device from $35\times50\,\mu\text{m}^2$ ("big") to $10\times20\,\mu\text{m}^2$ ("small"). Performing this comparison on the same single crystal ensures that the bulk mean free path is identical between the devices. By decreasing the contact separation from above ($35\,\mu\text{m} > l$) to below ($10\,\mu\text{m} < l$) the mean free path, we expect to see a qualitative change of the voltage signal in a field region where r_c is not the smallest length scale ($10\,\mu m < r_c \Rightarrow B < \frac{\hbar k_\text{F}}{e10\,\mu m} = \pm\, 0.65\,\text{T}$.)

The $PdCoO_2$ device before and after its size reduction is presented in Fig. 3.40a and b. The TEF spectra of both show characteristic peaks in positive fields from intra-nozzle focusing events, as well as from focusing between the top electrodes at negative fields. In the 'big' device the zero-field potential difference is positive as a result of the finite resistance due to bulk scattering processes. The 'small' device, on the other hand, shows a voltage inversion around zero-field. This is a ballistic effect which was well-studied and understood in four-probe junctions in semiconductor 2DEGs [37, 104]. Essentially electrons are ballistically projected from the current sourcing nozzle I^- to the voltage probing contact V^+. As more charge carriers arrive at V^+ than at the V^- nozzle, the observed voltage is negative.

I mention this experiment in particular in view of the voltage inversion recently observed in graphene [105]. The observed negative voltage in $PdCoO_2$ presented above shows that great care needs to be taken in order to exclude ballistic effects when discussing the often subtle features of hydrodynamic transport.

The transverse electron focusing experiments presented in this section showcase how rich and often counter-intuitive ballistic effects can be. We have discussed how the faceted Fermi surface of $PdCoO_2$ induces directional ballistic transport which can lead to a super-geometric enhancement of the ballistic properties. Moreover these experiments challenge our current microscopic understanding of ballistic electron transport and description of surface scattering. At the same time we have learned that in the purely ballistic regime, where $l \gg L$, simple ballistic models capture the qualitative and, for most parts, also quantitative behaviour of the TEF devices. With that, these experiments and simulations set the basis for future investigations.

References

1. Haynes WM (2014) CRC handbook of chemistry and physics, 95th edn
2. Daou R, Fresard R, Eyert V, Hebert S, Maignan A (2017) Unconventional aspects of electronic transport in delafossite oxides. Sci Technol Adv Mater 18:919–938
3. Tanaka M, Hasegawa M, Takei H (1996) Growth and anisotropic physical properties of $PdCoO_2$ single crystals. J Phys Soc Jpn 65(12):3973–3977
4. Takatsu H, Yonezawa S, Mouri S, Nakatsuji S, Tanaka K, Maeno Y (2007) Roles of high-frequency optical phonons in the physical properties of the conductive delafossite $PdCoO_2$. J Phys Soc Jpn 76:10
5. Kushwaha P, Sunko V, Moll PJW, Bawden L, Riley JM, Nandi N, Rosner H, Schmidt MP, Arnold F, Hassinger E, Kim TK, Hoesch M, Mackenzie AP, King PDC (2015) Nearly free electrons in a 5d delafossite oxide metal. Sci Adv 1(9)
6. Kushwaha P, Borrmann H, Khim S, Rosner H, Moll PJW, Sokolov DA, Sunko V, Grin Yu, Mackenzie AP (2017) Single crystal growth, structure, and electronic properties of metallic delafossite $PdRhO_2$. Cryst Growth Design 17(8):4144–4150
7. Mekata M, Sugino T, Oohara A, Oohara Y, Yoshizawa H (1995) Magnetic structure of antiferromagnetic $PdCrO_2$ possible degenerate helices on a rhombohedral lattice. Phys B 213:221–223
8. Takatsu H, Yonezawa S, Michioka C, Yoshimura K, Maeno Y (2010) Anisotropy in the magnetization and resistivity of the metallic triangular-lattice magnet $PdCrO_2$. J Phys: Conf Ser 200:012198
9. Coldea AI, Seabra L, McCollam A, Carrington A, Malone L, Bangura AF, Vignolles D, Van Rhee PG, McDonald RD, Sörgel RD, Jansen M, Shannon N, Coldea R (2014) Cascade of field-induced magnetic transitions in a frustrated antiferromagnetic metal. Phys Rev B 90(2)

10. Shin YJ, Doumerc JP, Dordor P, Pouchard M, Hagenmuller P (1993) Preparation and physical properties of the delafossite-type solid solutions $AgCo_xNi_{1-x}O_2$. J Solid State Chem 107:194–200
11. Shannon RD, Rogers DB, Prewitt CT (1971) Chemistry of noble metal oxides I. Syntheses and properties of ABO_2 delafossite compounds. Inorganic Chem 10(4):713–718
12. Prewitt CT, Shannon RD, Rogers DB (1971) Chemistry of noble metal oxides II. Crystal structures of $PtCoO_2$, $PdCoO_2$, $CuFeO_2$, and $AgFeO_2$. Inorganic Chem 10(4):719–723
13. Rogers DB, Shannon RD, Prewitt CT, Gillson JL (1971) Chemistry of noble metal oxides III. Electrical transport properties and crystal chemistry of ABO_2 compounds with the delafossite structure. Inorganic Chem 10(4):723–727
14. Tanaka M, Hasegawa M, Takei H (1997) Crystal growth of $PdCoO_2$, $PtCoO_2$ and their solid-solution with delafossite structure. J Cryst Growth 173(3–4):440–445
15. Mackenzie AP (2017) The properties of ultrapure delafossite metals. Rep Progress Phys 80(3):032501
16. Nandi N, Scaffidi T, Kushwaha P, Khim S, Barber ME, Sunko V, Mazzola F, King PDC, Rosner H, Moll PJW, König M, Moore JE, Hartnoll S, Mackenzie AP (2018) Unconventional magneto-transport in ultrapure $PdCoO_2$ and $PtCoO_2$. Nat Quantum Mater 3(66)
17. Mao ZQ, Maeno Y, Fukazawa H (2000) Crystal growth of Sr_2RuO_4. Mater Res Bull 35(11):1813–1824
18. Banszerus L, Schmitz M, Engels S, Goldsche M, Watanabe K, Taniguchi T, Beschoten B, Stampfer C (2016) Ballistic Transport Exceeding 28 μm in CVD Grown Graphene. Nano Lett (2):1387–1391
19. Harada T, Fujiwara K, Tsukazaki A (2018) Highly conductive $PdCoO_2$ ultrathin films for transparent electrodes. APL Mater 6(4):046107
20. Hasegawa M, Higuchi T, Tanaka M, Tsukamoto T, Shik S, Takei H (2001) Electronics structure of Delafossite-Type Metallic Oxide $PdCoO_2$. Mater Trans 42(6):961–964
21. Higuchi T, Tsukamoto T, Tanaka M, Ishii H, Kanai K, Tezuka Y, Shin S, Takei H (1998) Photoemission study on $PdCoO_2$. J Electr Spectrosc Related Phenom 92(1):71–75
22. Higuchi T, Hasegawa M, Tanaka M, Takei H, Shin S, Tsukamoto T (2004) Unoccupied electronic state of delafossite-type $PdCoO_2$ single crystal probed using inverse photoemission spectroscopy. Jpn J Appl Phys 43(2):699–700
23. Seshadri R, Felser C, Thieme K, Tremel W (1998) Metal-metal bonding and metallic behavior in some ABO_2 delafossites. Chem Mater 10:2189–2196
24. Tanaka M, Hasegawa M, Higuchi T, Tsukamoto T, Tezuka Y, Shin S, Takei H (1998) Origin of the metallic conductivity in $PdCoO_2$ with delafossite structure. Phys B 245(2):157–163
25. Noh H, Jeong J, Jeong J, Cho E, Kim S, Kim K, Min B, Kim H (2009) Anisotropic electric conductivity of delafossite $PdCoO_2$ studied by angle-resolved photoemission spectroscopy. Phys Rev Lett 102(25)
26. Kim K, Choi H, Min BI (2009) Fermi surface and surface electronic structure of delafossite $PdCoO_2$. Phys Rev B 80(3)
27. Eyert V, Frésard R, Maignan A (2008) On the metallic conductivity of the delafossites $PdCoO_2$ and $PtCoO_2$. Chem Mater **20**(6):2370-2373
28. Ong KP, Zhang J, Tse JS, Wu P (2010) Origin of anisotropy and metallic behavior in delafossite $PdCoO_2$. Phys Rev B 81(11)
29. Hicks CW, Gibbs AS, Mackenzie AP, Takatsu H, Maeno Y, Yelland EA (2012) Quantum oscillations and high carrier mobility in the delafossite $PdCoO_2$. Phys Rev Lett 109(11)
30. Kanter J (2015) Switchting magnetic phases by modifying the sodium vacancy ordering in Na_xCoO_2. PhD thesis, ETH Zurich
31. Usui H, Ochi M, Kitamura S, Oka T, Ogura D, Rosner H, Haverkort MW, Sunko V, King PDC, Mackenzie AP, Kuroki K (2019) Hidden kagome-lattice picture and origin of high conductivity in delafossite $PtCoO_2$. arXiv:1812.07213. Phys Rev Mater **3**:045002
32. Daou R, Frésard R, Hébert S, Maignan A (2015) Large anisotropic thermal conductivity of the intrinsically two- dimensional metallic oxide $PdCoO_2$. Phys Rev B 91(10)

33. Gusev GM, Levin AD, Levinson EV, Bakarov AK (2018) Viscous transport and Hall viscosity in a two-dimensional electron system. Phys Rev B 98(16)
34. Moll P, Kushwaha PJW, Nandi N, Schmidt B, Mackenzie AP (2016) Evidence for hydrodynamic electron flow in $PdCoO_2$. Science 351(6277):1061–1064
35. Ziman JM (1972) Principles of the theory of solids. Cambridge University Press, Cambridge
36. Hashimoto E, Ueda Y (2002) Electrical properties of ultra-high-purity metals. In: Purification process and characterization of ultra high purity metals, pp 249–276
37. Beenakker CWJ, van Houten H (1989) Billiard model of a ballistic multiprobe conductor. Phys Rev Lett 63(17):1857–1860
38. Beenakker CWJ, van Houten H (1991) Quantum transport in semiconductor nanostructures. Solid State Phys 44:1937–6871
39. Wen L, Roussel P, Pedreira O, Briggs B, Groven B, Dutta S, Popovici MI, Heylen N, Ciofi I, Vanstreels K, Østerberg FW, Hansen O, Petersen DH, Opsomer K, Detavernie C, Wilson CJ, van Elshocht S, Croes K, Bömmels J, Tökei Z, Adelmann C (2016) Atomic Layer Deposition of Ruthenium with TiN Interface for Sub- 10 nm Advanced Interconnects beyond Copper. ACS Appl Mater Inter 8(39):26119–26125
40. Sun T, Yao B, Warren AP, Barmak K, Toney MF, Peale RE, Coffey KR (2010) Surface and grain-boundary scattering in nanometric Cu films. Phys Rev B 81(15)
41. Rathmell AR, Nguyen M, Chi M, Wiley BJ (2012) Synthesis of oxidation-resistant cupronickel nanowires for transparent conducting nanowire networks. Nano Lett 12(6):3193–3199
42. Rakheja S, Chang S, Naeemi A (2013) Impact of dimensional scaling and size effects on spin transport in copper and aluminum interconnects. IEEE Trans Electr Devices 60(11):3913–3919
43. Maqableh MM, Huang X, Sung S, Reddy KSM, Norby G, Victora RH, Stadler BJH (2012) Low-resistivity 10 nm diameter magnetic sensors. Nano Lett 12(8):4102–4109
44. Datta S (1995) Electronic transport in mesoscopic systems. Cambridge University Press, Cambridge
45. Ihn T (2010) Semiconductor nanostructures. Cambridge University Press, Cambridge
46. Thompson JJ (1901) Proc Cambridge Philos Soc 11:120
47. Stone I (1898) On the electrical resistance of thin films. Phys Rev 6(1)
48. Fuchs K (1938) The conductivity of thin metallic films according to the electron theory of metals. Math Proc Cambridge Philos Soc 34(100):100–108
49. Sondheimer EH (1952) The mean free path of electrons in metals. Adv Phys 1(1)
50. Dingle RB (1950) The electrical conductivity of thin wires. Proc R Soc Lond Ser A, Math Phys Sci 201(1067):545–560
51. Soffer SB (1967) Statistical model for the size effect in electrical conduction. J Appl Phys 38(4):1710–1715
52. Chambers RG (1950) The Conductivity of Thin Wires in a Magnetic Field. Proc R Soc A: Math Phys Eng Sci 202(1070):378–394
53. MacDonald DKC, Sarginson K (1950) Size effect variation of the electrical conductivity of metals. Proc R Soc A: Math Phys Eng Sci 203(1073):223–240
54. Mayadas AF, Shatzkes M, Janak JF (1969) Electrical resistivity model for polycrystalline films: the case of specular reflection at external surfaces. Appl Phys Lett 14(11):345–347
55. Mayadas AF, Shatzkes M (1970) Electrical-resistivity model for polycrystalline films: the case of arbitrary reflection at external surfaces. Phys Rev B 1(4):1382–1389
56. Namba Y (1970) Resistivity and temperature coefficient of thin metal films with rough surface. Jpn J Appl Phys 9(11):1326–1329
57. Ashcroft NW, Mermin ND (1976) Solid state physics. Cengage Learning, Boston
58. Gaidukov YP (1984) Electronic properties of whiskers. Soviet Physics Uspekhi 27(4):256–272
59. Bass J (1972) Deviations from Matthiessen's rule. Adv Phys 21(91):431–604
60. Lee PA, Ramakrishnan TV (1985) Disordered electronic systems. Rev Modern Phys 57(2):287–337
61. Bergmann G (1984) Weak localisation in thin films. Phys Rep 107(1)

62. Lee PA, Stone AD, Fukuyama H (1987) Universal conductance fluctuations in metals: Effects of finite temperature, interactions, and magnetic field. Phys Rev B 35(3)
63. Trivedi N, Ashcroft NW (1988) Quantum size effects in transport properties of metallic films. Phys Rev B 38(17):12298–12309
64. Chandrasekhar V, Rooks MJ, Wind S, Prober DE (1985) Observation of Aharonov-Bohm electron interference effects with periods h/e and h/2e in individual micro-size, normal-metal rings. Phys Rev Lett 55(15)
65. Dauber J, Oellers M, Venn F, Epping A, Watanabe K, Taniguchi T, Hassler F, Stampfer C (2017) Aharonov-Bohm oscillations and magnetic focusing in ballistic graphene rings. Phys Rev B 96(20)
66. Landauer R (1957) Spatial variation of currents and fields due to localized scatterers. IBM J Res Develp 1:223–231
67. Büttiker M (1986) Four-Terminal Phase-coherent Conductance. Phys Rev Lett 57(14)
68. Onsager L (1931) Reciprocal relations in irreversible process II. Phys Rev 38:2265–2279
69. Casimir HBG (1945) On Onsager's principle of microscopic reversibility. Rev Modern Phys 17(2)
70. Englman R, Sondheimer EH (1956) The electrical conductivity of thin films. Proc Phys Soc 69:449–458
71. Ham FS, Mattis DC (1960) Electrical properties of thin-film semiconductors. IBM J Res Dev 1:143–151
72. Price PJ (1960) Anisotropic conduction in solids near surfaces. IBM J Res Dev 4(2):152–157
73. Parrott JE (1965) A new theory of the size effect in electrical conduction. Proc Phys Soc 85(6):1143–1155
74. Sato H, Yonemitsu K (1976) Anisotropic size effect in thin aluminium films. Phys Status Solidi B 73(2):723–733
75. Risnes R, Sollien V (1969) Anisotropy in the resistivity of thin aluminium films. Philosoph Mag 20(167):895–905
76. Hashimoto E, Ueda Y (1998) Anisotropy of the size effect in the electrical resistivity of high-purity Al single crystals. J Phys Condens Matter 10(30):6727–6734
77. Choi D, Moneck M, Liu X, Oh S, Kagan CR, Coffey KR, Barmak K (2013) Crystallographic anisotropy of the resistivity size effect in single crystal tungsten nanowires. Sci Rep 3
78. Zheng P, Gall D (2017) The anisotropic size effect of the electrical resistivity of metal thin films: Tungsten. J Appl Phys 122(13)
79. Walmsley P, Fisher IR (2017) Determination of the resistivity anisotropy of orthorhombic materials via transverse resistivity measurements. Rev Sci Inst 88(4)
80. Wu J, Bollinger AT, He X, Božović I (2017) Spontaneous breaking of rotational symmetry in copper oxide superconductors. Nature 547:432–435
81. Sharvin YV (1965) A possible method for studying Fermi surfaces. J Exper Theor Phys Lett 48:984–985
82. Goldman VJ, Su B, Jain JK (1994) Detection of composite fermions by magnetic focusing. Phys Rev Lett 72(13):2065–2068
83. Smet JH, Weiss D, Von Klitzing K, Fleischmann R, Ketzmerick R, Geisel T, Wegscheider W, Coleridge PT, Wasilewski ZW, Weimann G (1998) Composite fermions in magnetic focusing and commensurability experiments. Phys B 249–251:15–22
84. Yan C, Kumar S, Pepper M, See P, Farrer I, Ritchie D, Griffiths J, Jones G (2017) Temperature dependence of spin-split peaks in transverse electron focusing. Nanoscale Res Lett 12:10–15
85. Bladwell S, Sushkov OP (2018) Interference in spin-orbit coupled transverse magnetic focusing: emergent phase due to in-plane magnetic fields. Phys Rev B 98
86. Tsoi VS, Bass J, Wyder P (1999) Studying conduction-electron/interface interactions using transverse electron focusing. Rev Modern Phys 71:1641–1693
87. Tsoi VS (1974) Focusing of electrons in a metal by a transverse magnetic field. Pis'ma Zh Eksp Teor Fiz 19(2):114–116
88. Tsoi VS, Razgonov II (1976) Reflection of electrons and holes in antimony from the sample boundaries. Pis'ma Zh Eksp Teor Fiz 23(2):107–109

89. Tsoi VS, Razgonov II (1977) Specular reflection of conduction electrons from the crystal surface in tungsten or copper. Pis'ma Zh Eksp Teor Fiz 25(1):26–29
90. Tsoi VS, Bass J, Benistant PAM, van Kempen H, Payens ELM, Wyder P (1979) Transverse electron focusing and specular reflection in silver. J Phys F: Metal Phys 9(11):221–226
91. Sato H, Kimura F (1984) Transverse electron focusing in zinc single crystals. J Phys F: Metal Phys 14(8):1905–1918
92. Benistant PAM, van Kempen H, Wyder P (1985) Transverse electron focusing in aluminium. J Phys F: Metal Phys 15(12):2445–2450
93. Benistant PAM, van de Walle GFA, van Kempten H, Wyder P (1986) Comparison of transverse-electron-focusing and scanning- tunnelling-microscopy measurements on Ag(001) and (011) surfaces. Phys Rev B 33(2)
94. van Houten H, Beenakker CWJ, Williamson JG, Broekaart MEI, van Loosdrecht PHM, van Wees BJ, Mooij JE, Foxon CT, Harris JJ (1989) Coherent electron focusing with quantum point contacts in a two- dimensional electron gas. Phys Rev B 39(12):8556–8575
95. Dean CR, Young AF, Meric I, Lee C, Wang L, Sorgenfrei S, Watanabe K, Taniguchi T, Kim P, Shepard KL, Hone J (2010) Boron nitride substrates for high-quality graphene electronics. Nature Nanotechnol 5(10):722–726
96. Taychatanapat T, Watanabe K, Taniguchi T, Jarillo-Herrero P (2013) Electrically tunable transverse magnetic focusing in graphene. Nature Phys 9(4):225–229
97. Lee M, Wallbank JR, Gallagher P, Watanabe K, Taniguchi T, Fal'ko VII, Goldhaber-Gordon D (2016) Ballistic miniband conduction in a graphene superlattice. Science 353(6307):1526–1529
98. Rickhaus P, Makk P, Liu M, Richter K, Schönenberger C (2015) Gate tuneable beamsplitter in ballistic graphene. Appl Phys Lett 107:251901
99. Chen S, Han Z, Elahi MM, Habib KMM, Wang L, Wen B, Gao Y, Taniguchi T, Watanabe K, Hone J, Ghosh AW, Dean CR (2016) Electron optics with p-n junctions in ballistic graphene. Science 353(6307):1522–1525
100. Barnard AW, Hughes A, Sharpe AL, Watanabe K, Taniguchi T, Goldhaber-Gordon D (2017) Absorptive pinhole collimators for ballistic Dirac fermions in graphene. Nature Commun 8:15418
101. Song AM, Lorke A, Kriele A, Kotthaus JP (1998) Nonlinear electron transport in an asymmetric microjunction: a Ballistic rectifier. Phys Rev Lett 80(17):3831–3834
102. Hornsey RI, Marsh AM, Cleaver JRA, Ahmed H (1995) High-current ballistic transport through variable-width constrictions in a high-mobility two-dimensional electron gas. Phys Rev B 51(11)
103. Fleischmann R, Geisel T (2002) Mesoscopic Rectifiers Based on Ballistic Transport. Phys Rev Lett 89(1)
104. Baranger HU, Stone AD (1990) Geometrical effects on the Hall resistance in ballistic microstructures. Surf Sci 229:212–215
105. Bandurin DA, Torre I, Krishna Kumar R, Ben Shalom M, Tomadin A, Principi A, Auton GH, Khestanova E, Novoselov KS, Grigorieva IV, Ponomarenko LA, Geim AK, Polini M (2016) Negative local resistance caused by viscous electron backflow in graphene. Science 351:1055–1058

Chapter 4
Spatially Modulated Heavy Fermion Superconductivity in $CeIrIn_5$

In stark contrast to the nearly free electron system of $PdCoO_2$, strong electron interactions in $CeIrIn_5$ give rise to heavily renormalised quasiparticles at low temperatures. Moreover, $CeIrIn_5$ exhibits heavy fermion superconductivity below $T_c = 0.4\,K$. Due to the relatively low energy scale set by the hybridisation of the cerium $4f$ moments with the conduction electrons, the low temperature properties of $CeIrIn_5$ can readily be tuned by non-thermal parameters. In particular, the superconductivity is exceptionally sensitive to the application of uniaxial pressure, such that T_c can be increased by 100% with only 7 kbar.

Here I will demonstrate how the pressure sensitivity can be exploited to locally manipulate the electronic potential landscape and thereby spatially modulate the superconducting properties of $CeIrIn_5$. Only a few approaches have been attempted to gain control over the *local* ground state in strongly correlated metals, although electrons in engineered potentials are readily studied in low-dimensional semiconductor heterostructures and led to landmark discoveries. Encouraged by this, we have found a route to locally manipulate the strain within microstructured $CeIrIn_5$ devices. The approach is based on taking advantage of the biaxial strain induced by differential thermal contraction and spatially tailoring it by defined FIB cuts. In this manner, the local T_c within the lamella is controlled via the spatial strain distribution. This result presents a generic route to manipulate electronic order on micrometer length scales in strongly correlated matter.

The chapter begins with an overview of the basic concepts of heavy fermion physics, followed by an introduction to the intermetallic compound $CeIrIn_5$. Thereafter I will present my transport measurements of microstructured $CeIrIn_5$ devices, which led to the discovery of spatially modulated superconductivity. This observation puzzled us for a long time and led me to building up a body of evidence against a fully intrinsic mechanism explaining the data. Finally, finite element simulations and SQUID microscopy images of my lamellae, obtained by Katja Nowack's group at Cornell University, enabled us to clearly identify strain as the underlying cause for the spatially anisotropic superconductivity.

M. D. Bachmann, *Manipulating Anisotropic Transport and Superconductivity by Focused Ion Beam Microstructuring*, Springer Theses,
https://doi.org/10.1007/978-3-030-51362-7_4

4.1 Introduction to the Heavy Fermion Compound $CeIrIn_5$

This introduction begins with the concept of renormalised quasiparticles in the framework of Landau's Fermi liquid theory, which will clarify what exactly is 'heavy' about heavy fermion compounds. Thereafter I will present a general introduction to the heavy Fermion problem, leading to the discussion of Doniach's phase diagram, which summarises the essential physics of heavy fermion systems. And lastly I will introduce $CeIrIn_5$ and the closely related cerium-based heavy fermion compounds $CeCoIn_5$ and $CeRhIn_5$.

4.1.1 Interacting Electrons in Landau's Fermi Liquid Theory

In order to understand what is 'heavy' about $CeIrIn_5$, we must take a closer look at the electron-electron interactions in a metal. Unlike in a non-interacting Fermi gas, the electrons in metals interact with each other, which gives rise to electronic correlations. Mathematically, the description of the interacting system is straightforward to write down, however due to the large number of interacting particles ($N \sim 10^{23}$), the exact solution of the many body problem in a solid is impossible to calculate exactly. However, in the limiting case of low-energy excitations, the interacting electron problem was successfully addressed by Landau [1]. In his phenomenological theory, a system of interacting electrons, called the Fermi liquid, is described in terms of a Fermi gas of weakly interacting *quasiparticles* .

$$\begin{array}{ccc} \text{interacting Fermi gas of electrons} & \xrightarrow{\text{Landau's theory}} & \text{weakly interacting 'Fermi gas'} \\ \equiv \text{Fermi liquid} & & \text{of quasiparticles} \end{array}$$

In the transformed, weakly interacting system, the charge, spin, and momentum of the fermions remain unchanged, while the dynamical properties of the quasiparticles, such as their effective mass m^*, spin susceptibility, or compressibility, absorb the interactions of the system. Crucially, the transformation is performed adiabatically, such that there is a one-to-one correspondence between the single-particle excitations of the Fermi liquid and the quasiparticles (for sufficiently low energies). Consequently the discontinuity of the electron distribution function at the Fermi level E_F withstands the transformation and therefore, the adiabatic mapping retains the notion of particle-hole excitations across a well-defined Fermi surface. However, since the quasiparticles are not eigenstates of the interacting system, they are short lived and decay due to scattering processes. In fact, the concept of quasiparticles is only meaningful close to the Fermi surface E_F, where their lifetime τ is sufficiently long, as it decays as

$$\frac{1}{\tau} \propto (E - E_F)^2. \tag{4.1}$$

This implies that Fermi liquid theory is most powerful at low temperatures, since with enhanced temperatures, the quasiparticles' energies with respect to the Fermi level increase, $E - E_F \propto k_B T$, which shortens their lifetimes. Another key property of Fermi liquids and a direct consequence of (4.1) is a quadratic temperature dependence of the electron-electron scattering rate, which leads to the characteristic low-temperature resistivity behaviour of Fermi liquids in the case where Umklapp scattering is efficient

$$\rho(T) = \rho_0 + AT^2, \tag{4.2}$$

where ρ_0 is the residual resistivity from defect scattering. Equation (4.1) and conversely the T^2 behaviour of the resistivity can be qualitatively understood from a simple phase space argument: At a given temperature T, the phase space for the initial and final state of an electron to scatter from and into, scales with $\sim k_B T$, thus the total available phase space for the scattering event is proportional to $T \cdot T$.

Formally, the electron-electron interactions can be expressed by a phenomenological set of parameters known as the Landau parameters F_n^α, where $\alpha \in \{a, s\}$ denotes if F is symmetric or antisymmetric and n is a positive integer. The effective mass m^* of the quasiparticle, which is renormalised as it absorbs the electron interactions, can then be expressed in terms of the non-interacting band electron mass m_{band} and the symmetric Landau parameter F_1^s,

$$\frac{m^*}{m_{\text{band}}} = 1 + \frac{F_1^s}{3}. \tag{4.3}$$

Note that the effective mass m^* is defined as $\hbar k / v_g$, where v_g is the group velocity of the quasiparticle wave packets, $v_g = \hbar^{-1} \partial E / \partial k$ and the band mass m_{band} only includes the renormalisation due to the periodic lattice of the system and can be found by band-structure calculations. The effective mass in turn renormalises the density of states at the Fermi level, $N(E_F)$, and consequently all dependent properties, for instance the electronic specific heat C,

$$N(E_F) = \frac{m^* k_F}{\pi^2 \hbar^2} \quad \Rightarrow \quad \frac{C}{T} = \frac{m^* k_F}{3\hbar^2} k_B^2 T = \frac{m^*}{m_{\text{band}}} \frac{C_{\text{gas}}}{T} \equiv \gamma, \tag{4.4}$$

where k_B is the Boltzmann constant and C_{gas} is the electronic specific heat of the non-interacting Fermi gas, which is linear in temperature. Since the renormalised Sommerfeld coefficient γ and the effective mass m^* quantify the strength of the electron interaction and are experimentally accessible, they are often used as benchmark values to characterise the renormalisation effects in heavy fermion compounds. This clarifies that the 'heaviness' of heavy fermions is due to strong electron interactions, which lead to a large enhancement of the effective mass m^* of the quasiparticles. The question of why the electron interactions in heavy fermion compounds are so strong in the first place is the subject of the next section.

Landau's theory of Fermi liquids describes most correlated electron liquids with great success, even in cases of extremely strong interactions, for instance in heavy fermion systems with Sommerfeld coefficients as large as 8 J/molK^2 (in YbBiPt) [2], which is about 10^3 times higher than in non-interacting systems. This is a consequence of the fact that Fermi liquid theory is not a perturbative theory, which would be valid only for weak interactions. The relatively rare failure of Landau's theory, found in so-called non-Fermi liquids, is not thought to be associated with the strength of the electronic interaction, but rather with its range.

4.1.2 Heavy Fermion Systems and Kondo Physics

As mentioned above, an extreme situation showcasing the success of Landau's Fermi liquid theory can be observed in heavy fermion systems. These are found in intermetallic compounds containing either lanthanide (Ce, Yb) or actinide (U, Np) elements, and are characterised by partially filled $4f$ or $5f$ orbitals, which lead to the presence of localised magnetic moments. For instance, a trivalent cerium ion Ce^{3+} ($4f^1$), which can be found in $CeIrIn_5$, has a total angular momentum of $J = |L - S| = |3 - \frac{1}{2}| = \frac{5}{2}$ according to Hund's rules, and due to the small radius of the f-shells, which suppresses $f - f$ orbital overlap, localised magnetic moments are expected [3].

Indeed, the high-temperature, paramagnetic properties of heavy fermion systems are typically well-described by localised $4f$ moments on the rare-earth sites which weakly interact with the surrounding sea of conduction electrons [4]. However, as the temperature is lowered, the local moments progressively interact with the conduction electron system, which eventually leads to the formation of a non-magnetic singlet state. In this process, the f-electrons hybridise with the conduction electrons and become part of the Fermi sea, which enhances the density of states at the Fermi level accordingly. The increased itinerancy of the f-electrons is also evidenced by a cross-over from Curie–Weiss (local) to a Pauli (itinerant) behaviour of the susceptibility. It is this many-body entanglement between the f-electrons and the conduction electrons which leads to strong electron-electron interactions and gives rise to large effective masses.

The scenario described above is commonly known as the Kondo effect and was first described experimentally in the context of metals containing dilute magnetic impurities, for instance Au or Ag with traces of Fe, in the 1930s [5]. These alloys display an anomalous upturn in their low temperature resistivity, indicating the presence of an unconventional scattering mechanism at low temperatures. The phenomenon went unexplained until the 1960s, when Kondo understood that the anomalous scattering was linked the hybridisation of the localised impurity state with the conduction electrons. He investigated this scenario by studying the interaction between a localised impurity spin $\vec{S}$ coupled to a sea of conduction electrons via an exchange interaction J [6]. This led to the development of the so-called *sd* exchange model

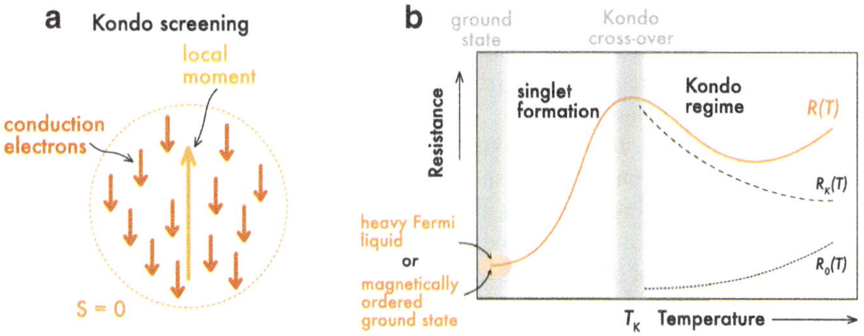

Fig. 4.1 **a** Schematic model of the Kondo screening. Below the T$_K$, the conduction electrons effectively screen the magnetic moment, which leads to a non-magnetic singlet ground state. **b** Characteristic temperature dependent resistance of a heavy fermion 'Kondo lattice' system. At elevated temperature the resistance increases as a function of temperature due to spin-flip scattering processes, while below a characteristic temperature T_K a non-magnetic Kondo singlet are formed

$$\mathcal{H}_{sd} \propto J\, \vec{s}(0) \cdot \vec{S} \tag{4.5}$$

where $\vec{s}(0)$ is the spin density of the conduction electrons at the position of the impurity. In this scenario a conduction electron can either interact with the localised spin $\vec{S}$ via ordinary potential scattering or through exchange, i.e. spin-flip scattering, in which a conduction electron exchanges its spin with the impurity. Using a perturbative approach in J for antiferromagnetic coupling ($J < 0$), Kondo found that in the lowest order of perturbation theory nothing special happens, however in second (and higher) order spin-flip processes add a logarithmically diverging term to the scattering amplitude and as a result the resistance exhibits a characteristic increase as the temperature is lowered

$$R(T) = R_0(T) + R_K\left(1 + 2|J|N(E_F)\ \ln \frac{E_F}{T}\right). \tag{4.6}$$

This form however diverges as a function of temperature, c.f. Fig. 4.1b. Taking all higher order spin-flip scattering terms into account, the critical temperature scale of the divergence is estimated to be

$$T_K \propto E_F\, e^{-1/|J|N(E_F)}. \tag{4.7}$$

Below T_K Kondo's perturbative model of strong spin-flip scattering is no longer valid. Instead, the localised spins are progressively screened by the conduction electrons, which reduces the overall scattering rate. In the case of antiferromagnetic coupling the conduction electrons with opposite spin to the impurity form a screening cloud and thereby create an overall non-magnetic ground state, c.f. Fig. 4.1a.

In a crystal such as CeIrIn$_5$, however, the magnetic impurities are densely spaced and hence the system can not be treated in the same dilute limit as originally studied by Kondo. Instead, Doniach advanced the idea of the so-called Kondo lattice, in which an array of weakly interacting magnetic moments are located periodically on sites of the lattice [7]. This transforms the *sd* Hamiltonian (4.5) into a form $J\ \sum_i \vec{s}_i(0)\cdot\vec{S}_i$, which sums over the lattice sites i.

In a Kondo lattice, there are two physical effects which influence the interaction between the conduction electrons and a lattice of localised moments. As discussed above, Kondo screening favours a non-magnetic Kondo-singlet ground state. Due to the strong hybridisation of the localised moments with the conduction electrons, the density of states at the Fermi level is increased and the Fermi liquid parameters are strongly renormalised, leading to a heavy fermion state. On the other hand, a magnetic moment immersed in a sea of conduction electrons can also polarise the local spin density, which results in a local modulation of the charge density $\rho(r)$. These so-called Friedel oscillations decay in an oscillatory fashion $\rho(r)\sim\frac{cos(2k_F r)}{|k_F r|^3}$ as a function of distance r away from the magnetic moment. This allows even strongly localised moments to interact with each other and promote a magnetically ordered ground state. This is known as the RKKY (Ruderman-Kittel-Kasuya-Yosida) interaction and favours antiferromagnetic or ferromagnetic coupling depending on the distance between the magnetic moments, c.f. Fig. 4.2a. The RKKY interaction leads to magnetic ordering below a characteristic temperature

$$T_{\mathrm{RKKY}}\propto N(E_{\mathrm{F}})J^2, \tag{4.8}$$

which, for typical interatomic distances relevant in heavy fermion materials (the in-plane Ce-Ce spacing in $CeIrIn_5$) favour antiferromagnetic coupling.

In summary, the physics of the heavy-fermion state is governed by the complex interplay between the non-magnetic Kondo singlet state and the long-range RKKY interaction favouring a magnetic ground state. This is commonly summarised in

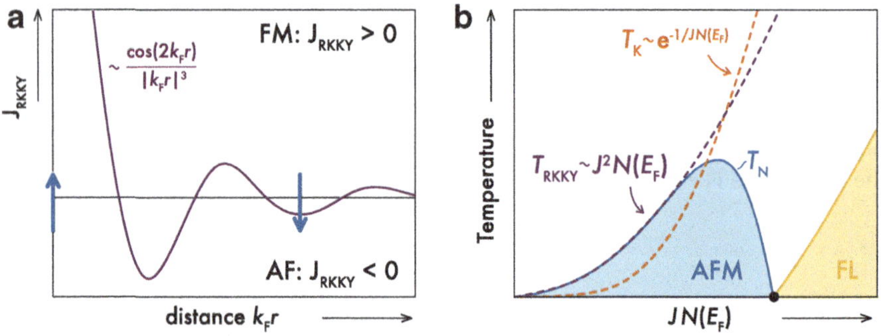

Fig. 4.2 **a** The magnitude of the RKKY interaction oscillates as a function of the dimensionless distance $k_F r$ away from the magnetic moment. Depending on the sign of the J_{RKKY}, the interaction can either be ferromagnetic (FM) for J_{RKKY} >0 or antiferromagnetic (AF) for J_{RKKY} <0. **b** The Doniach phase diagram explores the relative strengths of the RKKY and Kondo interaction as a function of $JN(E_F)$, which determines if the ground state of a heavy fermion system is magnetically ordered

the Doniach phase diagram [7], shown in Fig. 4.2b. For small couplings $|J|$, the RKKY energy scale is dominant and consequently a system in this regime orders magnetically below the Néel temperature T_N. As the hybridisation increases, the Kondo energy scale starts to increase. Consequently T_N is suppressed because the magnetic moments themselves are reduced due to Kondo screening, and eventually at large enough $|J|$, a paramagnetic heavy fermion liquid becomes the new ground state of the system. In Doniach's diagram, the ground states are separated by a quantum critical point, at which the magnetic order is suppressed to zero. At this point, quantum rather than thermal fluctuations govern the physical properties, which can lead to unusual phenomena such as non-Fermi liquid behaviour of exotic superconductivity, which I will discussion in the next section.

4.1.3 *Superconductivity in Heavy Fermion Systems*

Not long after the first heavy fermion compound $CeAl_3$ was found in 1975 [8], the first occurrence of superconductivity in the heavy fermion compound UBe_{13} was reported [9]. However, the measurement results of UBe_{13} left room for the possibility of non-intrinsic, filamentary superconductivity and were deemed inconclusive. The intrinsic nature of superconductivity in a heavy fermion compound was only convincingly shown in 1979, in measurements on $CeCu_2Si_2$ [10]. The discovery of superconductivity in the vicinity of strong magnetic interactions was unexpected and received a great deal of scientific interest as this presented the first possibility or exotic, magnetically-driven superconductivity.

To put this into context, I will discuss the key results of the microscopic theory of superconductivity introduced by Bardeen, Cooper and Schrieffer (BCS) in the following paragraphs. Intending not to drown in a sea of formulae, this brief review focuses on a qualitative description of conventional superconductivity and the comparison to other, exotic forms of superconducting behaviour. A detailed mathematical derivation can for instance be found in [11], which was used as a basis for this section.

The BCS theory of conventional superconductivity [12] is based on an instability of a Fermi gas against the formation of bound pairs of quasiparticles, so-called Cooper pairs. Cooper had shown that this instability can overcome the screened Coulomb repulsion between electrons and persist at sufficiently low temperatures if an arbitrarily weak *attractive* potential V between the electrons exists regardless of its origin [13]. Under these circumstances the binding energy of a Cooper pair is negative, i.e. the energy of a bound pair is lower than the combined energies of the non-interacting quasiparticles at the Fermi level [11].

The microscopic theory of BCS considers the special case of electron-phonon coupling, which causes a net electron-electron attraction via the excitation of a phonon and the associated lattice deformation, which locally creates a positive charge density. Under these conditions, a simple model can be constructed in which an attractive interaction potential $V_{k,k'}$ between two electrons with momentum $\vec{k}$ and $\vec{k'}$ exists in a small energy shell $\hbar\omega_c$ around the Fermi surface. Here ω_c defines the cut-off frequency, beyond which the electron interaction becomes repulsive. In the case of phonon mediated electron attraction, BCS equates ω_c with the Debye frequency ω_D, which characterises the phonon spectrum.

With this simple model the excitation spectrum of a superconductor can be calculated and one finds that the ground state energy of the Cooper pair condensate is separated by an energy gap 2Δ from the energy levels of the unpaired electrons. The temperature-dependent energy gap $\Delta(T)$ is a key aspect in the description of superconductivity and is ultimately the reason for the observed low temperature properties of a superconductor and accordingly dictates the superconducting critical temperature $k_\mathrm{B}T_c \sim \Delta(0)$. The gap can be calculated self-consistently using the so-called gap equation, which in the limiting case $T \to 0$ yields

$$\Delta(0) \approx 2\hbar\omega_\mathrm{D}\, e^{-\frac{1}{N(E_\mathrm{F})V}}. \tag{4.9}$$

In the limit of weak coupling $\lambda \equiv N(E_\mathrm{F})V \ll 1$, where $N(E_\mathrm{F})$ is the density of states at the Fermi level and V is the attractive pairing potential. The expression (4.9) highlights the relation between the three key factors $N(E_\mathrm{F})$, V, and ω_D, which determine the energy scale Δ of the superconducting condensate. The exponential factor depends on the inverse strength of the coupling and therefore exponentially suppresses the size of the gap for small values of $\lambda \ll 1$. In the limit of strong coupling $\lambda \gg 1$ on the other hand, the function $e^{-1/\lambda}$ saturates at the value of 1 and can therefore never diverge, even for diverging values of $N(E_\mathrm{F})$ or V. Attention should also be paid to the value of the Debye frequency ω_D, which directly influences the magnitude of the energy gap, as evidenced by isotope substitution experiments [14], which display $T_c \propto \omega_\mathrm{D} \propto \sqrt{1/M_\mathrm{ion}}$. These considerations led to the prediction of high temperature superconductivity in metallic hydrogen or hydrides, stimulating an experimental search which recently culminated in the demonstration of superconductivity at 203 K at high pressures in H_2S [15].

For the phonon-mediated case of BCS superconductivity, the simple functional form assumed for the attractive potential V implies a complete gap around the Fermi surface, which gives rise to the observed activated behaviour of the specific heat below the superconducting transition. However, electron-phonon interaction is not the only possible pairing mechanism and in general the energy gap Δ need not be isotropic. This can be the case if the pairing potential $V_{k,k'}$ is anisotropic to begin with, which may occur in non-phonon mediated superconductors. As the symmetry of the gap function $\Delta(\vec{k})$ is related to the symmetry of the superconducting wave function [16], it is instructive to investigate its possible symmetries.

The combined wave function of a Cooper pair consisting of two electrons with momentum k and k' and spin σ and σ' can be expressed as $\Psi(k, k', \sigma, \sigma')$. In the absence of spin-orbit coupling, Ψ can be separated into orbital Φ and spin χ degrees of freedom, $\Psi(k, k', \sigma, \sigma') = \Phi(k, k')\chi(\sigma, \sigma')$, which allows us to classify the superconducting condensate in terms of the orbital and spin symmetry. Due to Pauli's requirement of overall antisymmetry under the exchange of fermionic particles, $\Psi(k, k', \sigma, \sigma') = -\Psi(k', k, \sigma', \sigma)$, we find that the possible combinations of Φ

and χ are limited. For a symmetric, spin-singlet ($S = 0$) pairing, the orbital part is required to be antisymmetric and thus only allows for even orbital angular momenta $l = 0, 2, \ldots = s, d, \ldots$ and conversely for a spin-triplet state ($S = 1$), the orbital part is necessarily odd $l = 1, 3, \ldots = p, f, \ldots$

The consequences of the above analysis can be seen for instance in the high-temperature superconducting cuprates, which are nowadays widely believed to exhibit $l = 2$ (d-wave) pairing symmetry [17]. Additionally, some heavy fermion superconductors, such as $CeCoIn_5$ show strong evidence of d-wave pairing as well [18]. On the other hand, spin-triplet pairing (p-wave) is extremely rare. In most of these cases, the gap function vanishes along specific nodal directions and leads to power law dependencies of certain low-temperature properties such as the thermal conductivity or the specific heat, which, if persisting to sufficiently low temperatures, are a hallmark of unconventional superconductivity.

As heavy fermion systems are characterised by a large effective mass m^* and a high density of states at the Fermi level $N(E_F)$, superconducting heavy fermion systems are typically no longer in the weak, but rather in the strong coupling limit ($N(E_F)V \gg 1$). This suggests that the description of superconductivity in these systems may need to take into account effects beyond those of BCS, for instance exotic pairing mechanisms. Moreover, the large effective mass further implies that the Fermi temperature T_F is anomalously low (as $T_F \propto E_F = \frac{\hbar^2 k_F^2}{2m^*}$) and hence the energy scale T_c/T_F is comparable to that found in high temperature superconductors (HTSC). Adding to similarity with the HTSC, the superconducting dome in heavy fermion systems is often found in close proximity to a quantum critical point (QCP), where large quantum fluctuations are present [19]. This has provoked the idea of quantum critical fluctuations being the source of the attractive electron interaction; commonly discussed for instance is spin-fluctuation mediated superconductivity in the vicinity of a magnetic quantum critical point. In summary, the often close proximity of magnetism, quantum criticality and unconventional superconductivity in superconducting heavy fermion systems presents an excellent opportunity to study exotic superconductivity from which parallels may be drawn to the high temperature superconducting compounds.

4.1.4 $CeIrIn_5$ and Its Ce-Based '115' Relatives $CeCoIn_5$ and $CeRhIn_5$

The $CeMIn_5$ (M = Co, Rh, Ir) compounds, which are sometimes referred to as the cerium-based '115' family, were discovered in the early 2000s [20–22] and form a series of rare-earth intermetallic heavy-fermion systems. The electrons in these materials display a variety of correlated phenomena (c.f. Fig. 4.3), such as antiferromagnetism, unconventional superconductivity, heavy Fermi liquid or non-Fermi liquid behaviour. At the root of this behaviour are the $4f^1$ electrons of cerium, which exhibit aspects of both local and itinerant electron character. On the one hand,

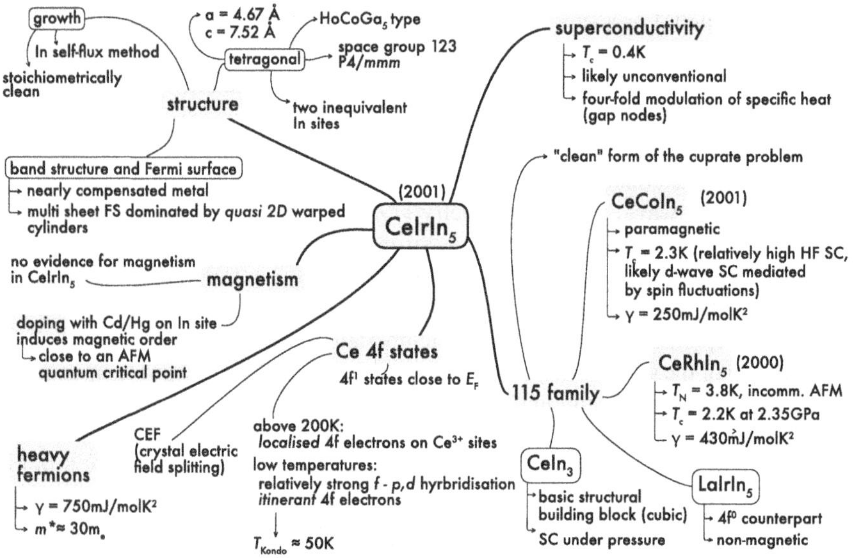

Fig. 4.3 Visual summary of the physical properties of $CeIrIn_5$. This mind map is intended as an overview and reminder of the characteristics of the 115 family. Abbreviations, detailed explanations, and references are provided in the main text

when sufficiently localised, the $4f$ state can lead to local moment magnetism with often complex magnetic order [23]. Alternatively, the $4f$ electrons can hybridise with the conduction electrons leading to strongly renormalised quasiparticles with a large mass enhancement, forming the so-called heavy fermion state. As the Kondo and RKKY interaction energy scales are comparable in magnitude, small perturbations induced for instance by moderate pressure, chemical doping or magnetic fields can readily favour one electronic state over another, c.f. Fig. 4.4. Moreover, the ground states are separated by quantum critical points, accompanied by quantum critical fluctuations, which in turn can give rise to correlated electronic states, such as unconventional superconductivity. Due to the close proximity of magnetism, quantum criticality and unconventional superconductivity in the 115 compounds, they are often regarded as prototypical materials for exploring the interplay of correlated states.

In the following I will present an overview of the physical properties of $CeIrIn_5$, which will elaborate on the mind map presented in Fig. 4.3. For the sake of clarity, I will focus on hydrostatic and uniaxial pressure as the non-thermal tuning parameters and will omit results from high field and chemical substitutions experiments for most parts. A recent review including these aspects can be found for instance in [27].

Beginning with the crystal structure, $CeIrIn_5$ is a member of the isostructural series $LMIn_5$ (L = La, Ce; M = Co, Rh, Ir), which crystallise in the tetragonal $HoCoGa_5$ form [28], displayed in Fig. 4.5b. The cerium (or lanthanum) atoms form a layered stack of square lattices, which are separated by iridium and indium atoms, giving

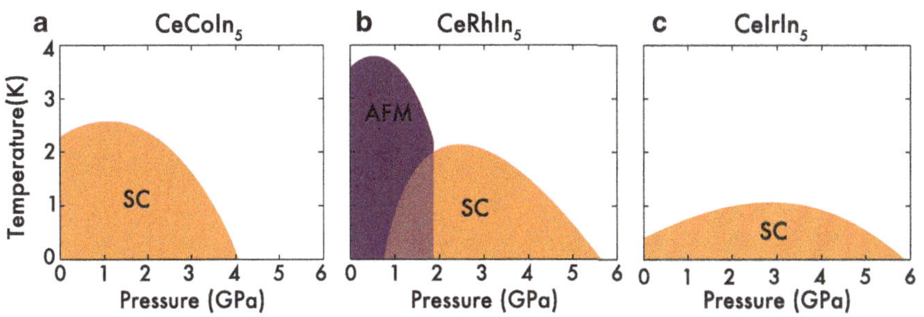

Fig. 4.4 The phase diagrams of CeCoIn$_5$ (**a**), CeRhIn$_5$ (**b**), and CeIrIn$_5$ (**c**) as a function of hydrostatic pressure are displayed on the same scale for a direct comparison. While CeIrIn$_5$ and CeCoIn$_5$ are ambient pressure superconductors, CeRhIn$_5$ can be tuned from an antiferromagnetic ground state to an unconventional superconductor by applying moderate hydrostatic pressures. Diagrams adapted from [24–26] respectively

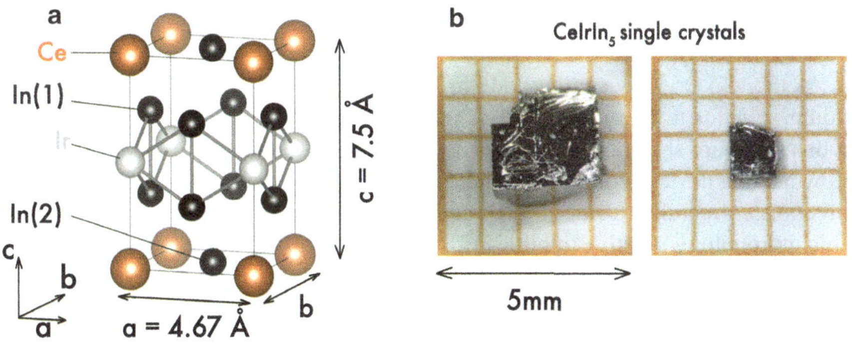

Fig. 4.5 **a** CeIrIn$_5$ crystallises as a tetragonal structure with lattice constants a = b = 4.67 Å and c = 7.5 Å. The layered structure hosts to inequivalent indium sites. **b** Single crystals of CeIrIn$_5$ prepared using an indium flux method. The sides of the as-grown crystals are parallel to the principal axes, such that an alignment of the crystal in the FIB is straightforward when cutting a lamella

rise to two inequivalent indium sites, one in the plane of the cerium atoms, In(2), and one in between the cerium layers, In(1). Occasionally the crystal structure is also described in terms of stacked layers of cubic CeIn$_3$ building blocks, separated by layers of IrIn$_2$. This is done in view of the theme of reduced dimensionality, which (incorrectly) hypothesises that higher superconducting transition temperatures are enabled in systems with lower dimensionality. In this scenario CeIrIn$_5$ is portrayed as the quasi two-dimensional relative of the three-dimensional, cubic compound CeIn$_3$, which becomes superconducting under pressure with a maximal T_c of 190 mK [19]. However, the next compounds in the general Ce$_n$$M_mIn_{3n+2m}$ series do not confirm this trend [27]. This discussion aside, the lanthanum containing compounds are the non-magnetic and non-superconducting $4f^0$ counterparts of the cerium-based 115s, and serve as a reference from which to infer the effect of the $4f$ moments, c.f. Fig. 4.7a for example.

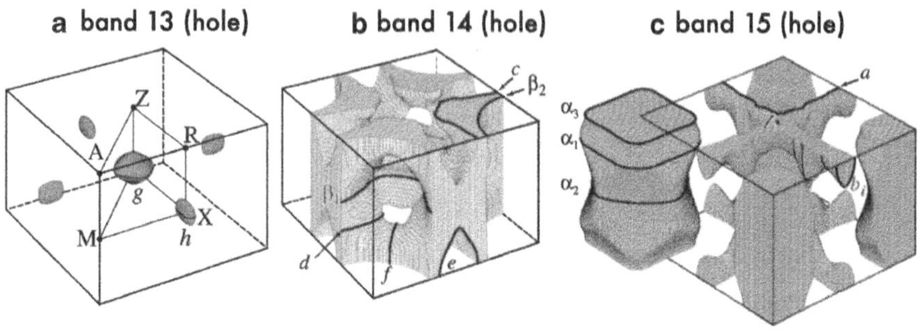

Fig. 4.6 Calculated Fermi surface sheets of $CeIrIn_5$. The orbits attributed to the measured dHvA frequencies are labeled with greek letters. While band 13 gives rise to fairly three-dimensional hole pockets, bands 14 and 15 bring about large, two-dimensional cylinders with large effective electron masses. Note that the calculated bands 1–12 do not cross the Fermi level as they lie lower in energy and hence do not contribute to the electronic properties of $CeIrIn_5$. Reprinted figure with permission from [30], Copyright (2001) by the American Physical Society

As the 115 crystals contain indium, they can conveniently be grown using an indium self-flux technique [29]. Two examples of $CeIrIn_5$ crystals from batches used in this study are shown in Fig. 4.5a. They exhibit the commonly observed columnar growth along the tetragonal *c* axis and furthermore the principal axes of the in-plane direction are parallel to the growth edges, as verified by Laue X-ray diffraction. This morphology greatly simplifies FIB microstructuring, as the orientation of the crystal can easily be re-identified in the FIB chamber. The crystal on the left shows some remaining indium flux on the bottom right corner of the top surface, which can be avoided when cutting lamellae. Lastly, the cerium-based 115 compounds are all stable in air, unlike $CeIn_3$, which further simplifies their handling during fabrication and subsequent measurements.

Band-structure calculations [30], presented in Fig. 4.6, provide a rough idea of the electronic structure of $CeIrIn_5$. The Fermi surface is comprised of multiple sheets, which give rise to a rich quantum oscillation spectrum (c.f. Fig. 4.15). While the effective electron masses associated with the relatively small and three-dimensional sheets from band 13 are comparatively light $\sim 6\,m_0$, the warped cylinders from the calculated bands 14 and 15 give rise to heavy quasi particles with masses up to $30\,m_0$.

Additionally, the size and topology of the Fermi surface are related to the hybridisation of the f electrons with the conduction band and are therefore indicative of the itinerancy of the $4f$ states. Haga et al. [30] conclude from their dHvA measurements that Fermi surface of $CeIrIn_5$ can be explained by largely itinerant $4f$ electrons. This is in line with the large value of the Sommerfeld coefficient γ determined from specific heat measurements, which yields $\gamma_{CeIrIn_5} = 750\,\mathrm{mJ/molK^2}$ [21]. Moreover, the hybridisation of the $4f$ states in $CeIrIn_5$ is notably stronger than in the related 115 compounds with respective coefficients $\gamma_{CeCoIn_5} = 250\,\mathrm{mJ/molK^2}$ [22] and $\gamma_{CeRhIn_5} = 430\,\mathrm{mJ/molK^2}$ [20].

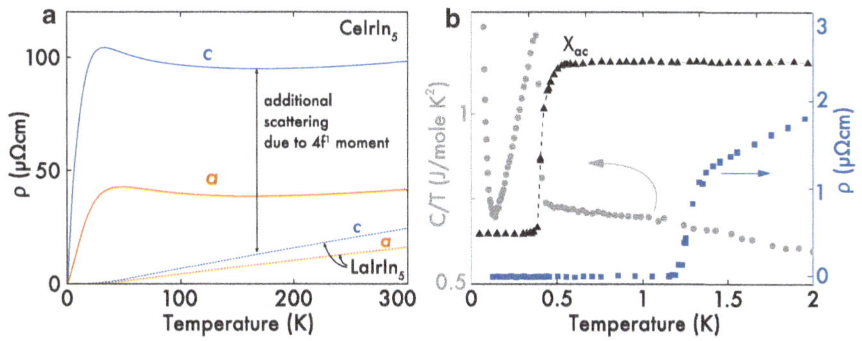

Fig. 4.7 **a** The temperature dependent resistivity of a CeIrIn$_5$ microstructure yields a resistivity anisotropy of approximately two at high temperatures and reveals a Kondo coherence peak around $T_K = 50$ K. The difference with the corresponding measurement of a microstructured LaIrIn$_5$ device emphasises the effect of the $4f^1$ moment from the cerium. **b** Measurements of thermodynamic quantities such as the specific heat C and the ac susceptibility χ_{ac} demonstrate that CeIrIn$_5$ undergoes a superconducting transition at $T_c = 0.38$ K. At the same time the resistivity shows an anomalously high temperature transition into a zero-resistance state at approximately 1.2 K. Figure reproduced from [21]

Next, I will review the transport properties of CeIrIn$_5$ by examining the temperature dependent resistivity. The direct comparison between a microstructured CeIrIn$_5$ device (solid curves) and a LaIrIn$_5$ structure (dashed curves) in Fig. 4.7a clearly shows the magnetic contribution of the $4f$ moments. Characteristic for a cerium-based heavy fermion compound, the resistivity of CeIrIn$_5$ at high temperatures is dominated by strong, incoherent scattering of the conduction electrons with the localised $4f$ moments. The local character of the f states at high temperature is further evidenced by the value of the effective magnetic moment, which is close to the Hund's rule value for Ce^{3+} [21]. As the temperature decreases, the conduction electrons progressively screen the magnetic moment, leading to a resistivity maximum around 50 K, referred to as the Kondo coherence peak. The maximum is associated with the development of correlated bands as the $4f$ electrons hybridise with the conduction electrons and gain coherence, which leads to a substantial reduction of the resistivity. The residual resistivity ratio (RRR) obtained in the microstructures devices is around 40, evidencing the high quality single crystal growth of CeIrIn$_5$.

At low temperatures, CeIrIn$_5$ undergoes a superconducting transition, c.f. Fig. 4.7b. Intriguingly, however, the transition into a zero-resistance state around 1.2 K does not coincide with the superconducting transition temperature $T_c = 0.38$ K determined by measurements of thermodynamic quantities such as the specific heat or the magnetic susceptibility. This anomaly has been subject to numerous experimental investigations and is commonly attributed to either an intrinsic form of filamentary superconductivity [31] or a competing precursor state to unconventional superconductivity [32]. In fact, the anomalous zero-resistance transition plays a significant role in this thesis as well, and will be shown to originate from surface damage-induced strain due to wire sawing.

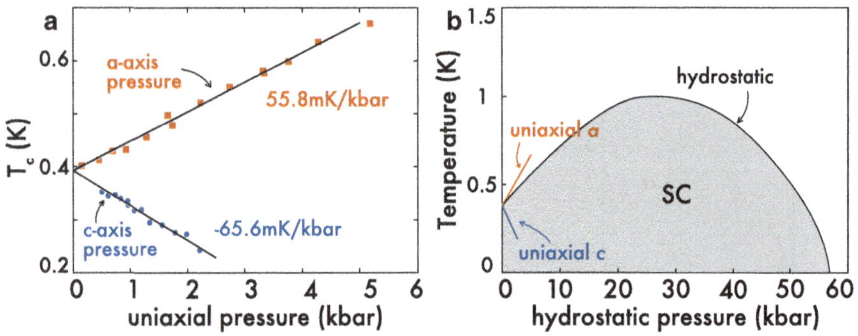

Fig. 4.8 **a** Dependence of the superconducting transition temperature T_c on uniaxial strain along the *a* and *c*-axis respectively. Figure reproduced from [42]. **b** Hydrostatic (isotropic) pressure dependence of T_c in $CeIrIn_5$. Reproduced from [26]. The uniaxial results from panel (A) are overlaid for a direct comparison

The superconducting state in the cerium-based 115 compounds has been most extensively investigated in $CeCoIn_5$, as it displays the highest T_c amongst the three compounds, c.f. Fig. 4.4. There is convincing evidence that the superconducting energy gap in $CeCoIn_5$ has line nodes, as demonstrated by the power-law (rather than exponential) temperature dependence of the thermal conductivity, specific heat [33, 34] and spin relaxation rate [35] below T_c. Moreover, scanning tunnelling microscopy further evidences the $d_{x^2-y^2}$ symmetry of the gap in $CeCoIn_5$ [36, 37]. Comparable power laws are observed in $CeIrIn_5$ [38, 39] and $CeRhIn_5$ [40], which suggests the presence of a similar unconventional superconducting state. Further, the emergence of superconductivity near T_N in $CeRhIn_5$ promotes the idea of magnetic fluctuation mediated Cooper-pairing in these systems [19, 41].

Moreover, by applying a hydrostatic pressure, the superconducting transition in $CeIrIn_5$ can be increased to 1.2 K, c.f. Fig. 4.8b. The change of T_c is even more dramatic if the system is subject to uniaxial stress as displayed in Fig. 4.8a. This high sensitivity can be understood in the context of the hybridisation of the $4f$ electrons, which is sensitive to the in-plane Ce-Ce interatomic distance. By compressing the sample along the *c* axis, the Ce-Ce distance increases proportionally to Poisson's ratio, which causes a linear decrease of T_c. Conversely compression along the *a* axis causes an almost equal enhancement of the superconducting transition temperature temperature. The application of a uniform compression also decreases the Ce-Ce distance and therefore the $4f$ hybridisation, albeit not as efficiently uniaxial strains do, as demonstrated with the slower increase of T_c under hydrostatic pressure. The high sensitivity and anisotropic response of $CeIrIn_5$ to uniaxial strain make the material an ideal model system to examine the effects of biaxial strain and investigate the possibility of spatially modulating the $4f$ hybridisation across a device.

In summary, $CeIrIn_5$ is a characteristic cerium-based heavy fermion superconductor. The hybridisation of the $4f^1$ electrons with the conduction band is comparatively strong as evidenced by large Sommerfeld coefficient and heavy effective electron masses of up to $30\,m_0$. Further, dHvA measurements demonstrate that the heavy quasiparticles originate from large, quasi two-dimensional Fermi surface sheets. Finally, the superconductivity observed at ambient pressures is dependent on the hybridisation of the $4f$ electrons and is thus highly sensitive to applied strains.

4.2 Electrical Transport Investigations

Often electrical transport measurements are one of the first experiments performed to probe the temperature, magnetic field or pressure dependent properties of a material. Compared to many other techniques, a resistivity measurement is set up relatively quickly, but can be less straightforward to interpret. This is partly because the electrical resistivity of a material does not directly relate to the thermodynamic state of system, but is a transport property instead. As such, the resistivity describes the response of a conductor to an external electric field, which is balanced against internal scattering effects of the charge carriers. In addition to this, a standard resistivity measurement determines the macroscopic average of a bulk sample. Microscopic details, for instance inclusions, crystal imperfections, domain boundaries, or local chemical potential variations influence and contribute to the resistivity but can be hard to identify using transport probes alone. Hence resistivity data should be interpreted with care and considerable attention should be paid to details.

With the above in mind I will present a series of resistivity measurements of microstructured $CeIrIn_5$ lamellae in the following section. Much to our surprise, we found that the superconducting transition into a zero-resistance state, when sourcing the current along the a or c axis respectively varied up to a factor $\frac{T_c^c}{T_a^c} = \frac{1.2\,\mathrm{K}}{0.25\,\mathrm{K}} \approx 5$. Using a combination of zero-field and in-field transport measurements, innovative sample designs, and critical current and quantum oscillation measurements, we could identify the underlying reason for the split superconducting transition to be the result of biaxial substrate strain, which induces a highly non-trivial strain pattern in the microstructured samples. This finding led us to re-examine the long-standing mystery of the anomalously high resistive transition in macroscopic samples at $T_c^* \approx 1.2\,\mathrm{K}$ [21], which is about three times higher than the reported bulk superconducting transition at $T_c^{\mathrm{bulk}} \approx 0.4\,\mathrm{K}$ (c.f. Fig. 4.4a). Our investigations demonstrate that surface damage induced strain is the reason for the observed resistive behaviour. Finally, and with the newly gained insights, we have produced bulk samples that show a resistive transition at $T_c^{\mathrm{bulk}} = 0.38\,\mathrm{K}$ for the first time reported.

4.2.1 *Superconducting Transition in Zero Field*

In microstructured samples of $CeIrIn_5$, the superconducting transition temperature along the c axis is consistently higher than the thermodynamic bulk value of

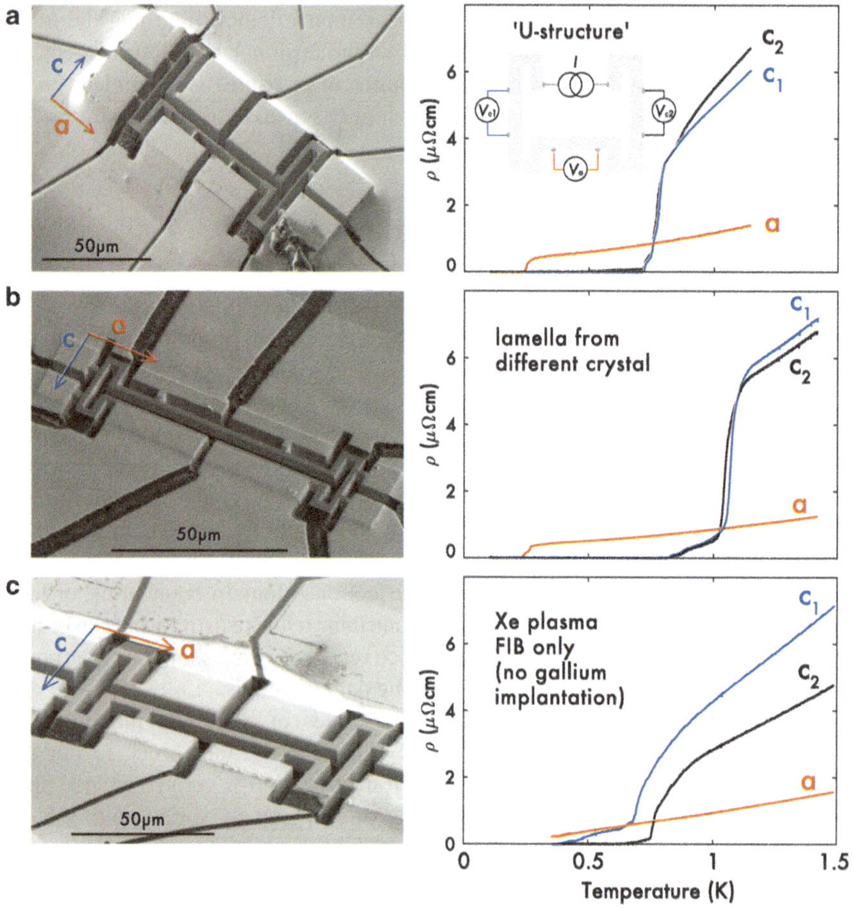

Fig. 4.9 **a** The temperature dependent resistivity along of a microstructured $CeIrIn_5$ lamella (left) shows a pronounced difference between the superconducting transition along the a and c axis. The inset shows a typical sample design we call 'U-structure', which allows us to simultaneously measure transport along two crystallographic directions. **b** A lamella cut from a different crystal reproduces the result of a split superconducting transition. **c** A lamella cut and structured exclusively in the xenon plasma FIB demonstrates that the split transition is not a result of gallium implantation

$T_c^{\text{bulk}} = 0.38\,\text{K}$, and does not occur simultaneously with the transition along the a axis. This behaviour is highly unusual, as the superconducting state is a macroscopically coherent phenomenon, which affects the entire sample at once and hence should occur isotropically. In Figs. 4.9 and 4.10, I compile a series of zero-field measurements of numerous samples, which provides a first insight into this unexpected observation.

Most of the microstructured devices were milled into 'U' shaped structures (c.f. inset in Fig. 4.9a), which permit the measurement of two c axis bars and one

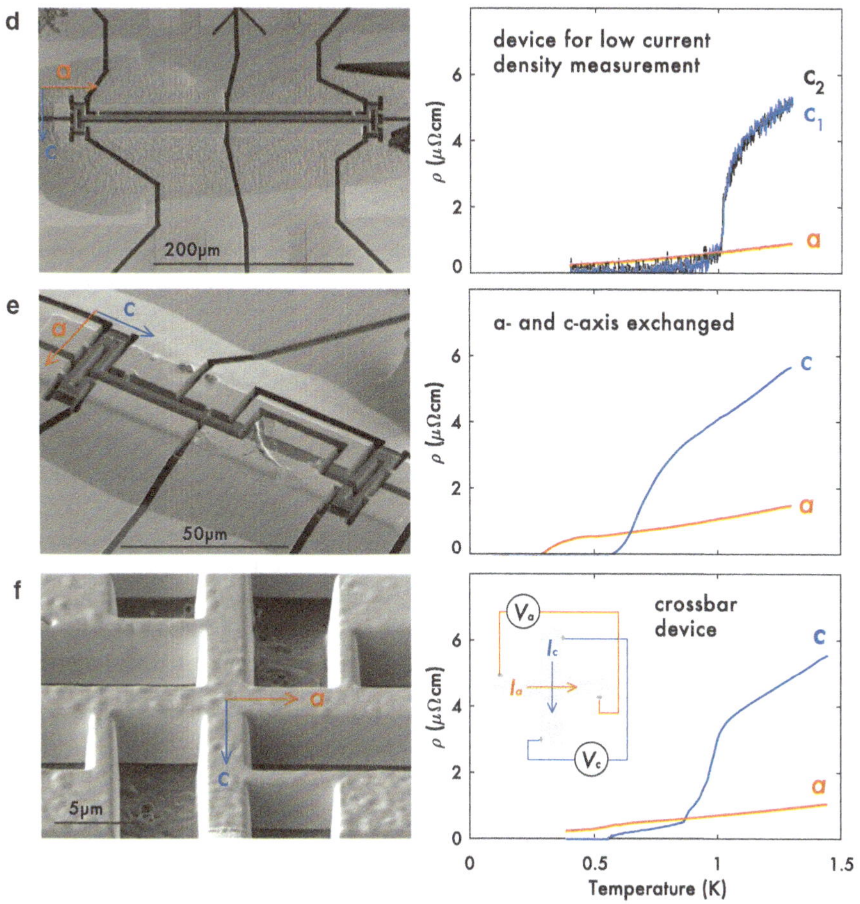

Fig. 4.10 **d** An extremely long lamella was fabricated in order to measure with low current density. Remarkably the two *c* axis bars, which are separated by over 200 μm, show the exact same temperature behaviour rendering a scenario evolving local defects an unlikely cause of the elevated transition temperature. **e** To exclude systematic effects from microstructuring, a sample was fabricated with its *c* axis parallel to the long edge of the lamella. **f** In order to probe the exact same region of a device a cross-bar structure was sculpted (see inset), to enable simultaneous measurement of the resistivity along the *a* and *c* axis

a axis bar simultaneously (or alternatively two *a* bars and one *c* bar, as shown in panel e). Additionally, the structure shown in panel f was milled into a 'crossbar' device. Here, beams along the *a* and *c* pass through each other, such that the central region of the cross can be simultaneously probed along both directions. Based on the measurement of the microstructured devices presented in Figs. 4.9 and 4.10 we conclude that

- the superconducting transition temperature along *c* is always higher than along *a*, strongly indicating a systematic underlying cause.

- T_c^a and T_c^c vary between samples, but are consistent within one lamella, even when the voltage probes are separated by over 220 μm, such as in panel d. This suggests that T_c^a and T_c^c are in some way externally influenced, for instance by the chosen geometry, and not solely set by the internal composition of the crystal.
- the residual resistivity ratio $\frac{R(300K)}{R(2K)} = 17$ is in quantitative agreement with measurements on macroscopic crystals [43], indicating that no significant crystal defects were introduced during the fabrication process.
- panel b: to rule out any crystal dependence, we have used samples from multiple batches. The device shown in panel b was fabricated from a crystal grown over ten years ago, and stored since then in air. Its transport characteristics are consistent with three newly grown crystals from two different growth batches.
- panel c: microstructures fabricated exclusively with a xenon FIB show no qualitative difference in their superconducting behaviour. Thus, gallium doping or a chemical reaction at the sample surface forming a gallium alloy, which can occasionally induce surface superconductivity [44], can be excluded as the origin of the increased transition temperature.
- panel d: typically, the current density used to measure microstructures is higher than that used for macroscopic samples as a result of the vastly different cross sections. To exclude a scenario in which T_c^a is suppressed due to the use of a current density j exceeding the critical current density j_c^a, a 'low current density' device was fabricated. It consists of a 220 μm long transport bar along the a direction, which generates a high device resistance and hence a lower current can be sourced while maintaining a reasonable signal-to-noise ratio. In this device the current density was reduced to less than 0.5 A/cm^2, four times lower than that used for bulk measurements [45]. The zero-resistance transition along a, however, remains much lower than along the c direction.
- panel e: devices in panel a-d were all carved from a face orthogonal to the c direction, which therefore is always parallel to the short edge of the lamella. This opens the possibility of systematic fabrication artefacts. To this end, the device in panel e was cut from a face orthogonal to the a axis, thus reversing the direction of all fabrication steps and of the axis in the microstructure. Yet the transport results remain unchanged and we can conclude that the transport phenomenology perfectly follows the crystal axes and not the device orientation.
- panel f: finally, the crossbar device displayed in panel f indicates that, surprisingly, the resistivity along the a direction is only marginally influenced by the simultaneously existing zero-resistance state along the c direction.

In summary, the observed behaviour is unlikely to be due to FIB damage or fabrication and is reproducible in many devices from numerous crystals. To further investigate this curious phenomenon of split zero-resistance transitions, I fabricated two additional devices, which directly probe the resistivity anisotropy. The first consists of a traditional Montgomery geometry, while the second takes the form of a semi-circle, which both will be presented in the next section.

4.2.2 Transport Geometries for Anisotropic Conductors

Before the advent of modern fabrication techniques, electrical transport measurements of three-dimensional samples were primarily performed on rectangular prism shaped transport bars [46]. For anisotropic compounds, where the resistivity takes the form of a symmetric rank 2 tensor, one could either cut up to three transport bars oriented along the respective principal directions of the crystal, or use Montgomery's method [47] to directly determine the resistivity anisotropy in a single, rectangular sample. While the latter has the drawback of commonly yielding small voltage response signals, it bears the key advantage that the resistivity tensor can be determined within one well-defined volume of a crystal. This guarantees that the microscopic details between the axes are identical when determining the resistivity anisotropy. For this reason, I performed a measurement of the resistivity anisotropy in $CeIrIn_5$ in a Montgomery geometry, which I will present in the following section.

The Montgomery method, which is the generalised anisotropic version of the isotropic van der Pauw set-up [48], is based on a rectangular prism sample geometry with point-like contacts at the corners. Without loss of generality, I will discuss the case of a tetragonal crystal system (i.e. $CeIrIn_5$), in which the resistivity components ρ_a and ρ_c are sought. Crucially, for the Montgomery method to be treated as a two-dimensional problem, we ensure that the current is injected homogeneously in the out-of-plane direction and therefore only flows within the plane of the sample. With the geometry presented in Fig. 4.11a, one can then perform two independent resistance measurements by sourcing a current between two adjacent contacts and measuring the voltage drop between the remaining pair. This yields two resistance curves as a function of temperature, $R_1(T)$ and $R_2(T)$, which both depend on the value of $\rho_a(T)$ and $\rho_c(T)$, c.f. Fig. 4.11b. The task is now to find the two functions f_1 and f_2, which map the resistances in the known geometry onto the resistivities. Formally, this can be expressed as

$$\begin{aligned} \rho_a &= f_1(R_1, R_2, l, w, t) \\ \rho_c &= f_2(R_1, R_2, l, w, t), \end{aligned} \tag{4.10}$$

where l, w, and t are the length, width, and thickness of the device respectively. Montgomery [47] and later Dos Santos et al. [49] have given graphical and analytical solutions for this problem for selected sample geometries. However, in general these functions are determined by the solution of Laplace's equation in the specific geometry of the sample.

The sample in Fig. 4.11 has contacts of finite width and has rounded corners despite the high precision FIB machining, such that I chose to conduct a finite element calculation to solve the Laplace problem in the true sample geometry. For that I used the AC/DC module of the commercial finite-element software package COMSOL [50] to calculate the electric potential distribution shown in panel Fig. 4.11c. As the current and voltage contacts run along the entire thickness of the crystal, an out-of-plane current flow can be ignored and it is sufficient to study a stationary, 2D model

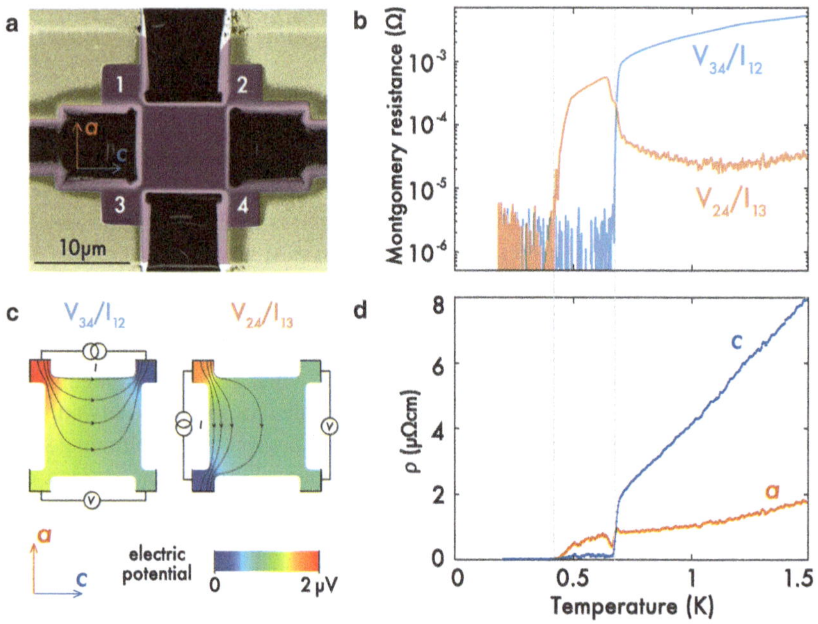

Fig. 4.11 **a** A Montgomery transport device consists of a central square with contacts in every corner. The current is sourced between adjacent corners, while the voltage is measured between the remaining two contacts. The lateral dimensions of the square are 10x10 μm^2 and the thickness is 6 μm. **b** The non-local Montgomery resistances as a function of temperature. **c** Finite element simulations explain the observed two orders of magnitude difference of the Montgomery resistance due to the of the normal state resistance anisotropy of approximately a factor 4. **d** From the data in panel b, the a and c axis resistivity can be extracted based on a numerical solution of the Laplace equation for the geometry in panel a

within the 'Electric Currents (ec)' interface. The only further necessary assumption is homogeneity of the material, i.e. that the device can described by a global resistivity tensor $\rho = \begin{pmatrix} \rho_a & 0 \\ 0 & \rho_c \end{pmatrix}$. (Note that, as mentioned above, we have restricted the problem to a 2D treatment and can omit the out-of-plane components.) With this and knowing the geometry of the sample, it is straightforward to solve the inverse problem of (4.11), i.e. to find the mapping

$$\begin{aligned} R_1 &= \tilde{f}_1(\rho_a, \rho_c, l, w, t) \\ R_2 &= \tilde{f}_2(\rho_a, \rho_c, l, w, t). \end{aligned} \tag{4.11}$$

This can directly be implemented in COMSOL for a dense array of realistic parameters of ρ_a and ρ_c, and the measured values for l, w and t. The result is a discrete mapping of $(\rho_a,\rho_c) \rightarrow (R_1,R_2)$. Lastly, the resistivities of the sample are found by numerically inverting the mapping function to find $(R_1,R_2) \rightarrow (\rho_a,\rho_c)$. In practice this can be implemented as an optimisation problem which finds the simulated pair (ρ_a,ρ_c)

that yields the closest value of (R_1,R_2). On a final note, the results obtained in this numerical fashion are excellently reproduced by directly solving two-dimensional Montgomery problem for a square geometry [47].

The resulting resistivities are presented in panel (4.10)d. Above 0.7 K the resistivity reproduces the reported anisotropy of $\rho_c/\rho_a \approx 4$. At approximately 0.7 K, the c axis shows a sharp transition into a low resistance state (numerical artefacts cause c to remain finite), while the resistivity along a remains mostly unchanged (the dip at 0.7 K is again due to a numerical artefact). Only once the temperature drops below $T_c = 0.4$ K is an isotropic zero-resistance state revealed. In essence, this result confirms the measurement of the crossbar device presented in Fig. 4.10f, which once more raises the issue: How can a device seemingly superconduct in one, but not the other direction?

To address this question, one may ask how T_c evolves as a function of angle θ away from the c axis, towards the a axis. This can be determined by performing a resistance measurement away from the principal axes of the lamella. For a current flowing at an angle θ away from the c axis, we expect the total measured resistivity to as $\rho_{\text{tot}} = \rho_a \sin^2\theta + \rho_c \cos^2\theta$ in the normal state, according to Eq. (3.9). This decomposition allows us think about an off-axis resistance measurement as two resistors in series with respective resistivities ρ_a and ρ_c. If there were a distinct mechanism leading to an enhancement of T_c exclusively along the c axis, one might qualitatively expect a partial transition at T_c^c, shorting out the ρ_c serial resistor contribution, followed by a transition into a zero-resistance state at T_c^a, as sketched in Fig. 4.12a. The situation, however, is not quite as straightforward, since the conductivity matrix in Eq. (3.9) becomes non-invertible as σ_c diverges. Nevertheless, in a scenario with only two intrinsic values of T_c, one may still not expect any intermediate values to occur.

To this end, the angular dependence of the superconducting transition is directly probed along a semi-circular current path within one microstructure, displayed in panel Fig. 4.12b (in purple). This sample demonstrates that FIB microstructuring not only facilitates the fabrication of complex device designs on microscopic length scales –it renders the fabrication of semi-circular samples possible in the first place, as it is impracticable to produce curved current paths with traditional fabrication techniques, such as wire-sawing, spark-cutting or polishing. In this microstructure, the central gold top contacts (large yellow pads) are the current source and sink, while the outward facing contacts form voltage pairs. The device is designed symmetrically around the central vertical axis, which allows for a systematic comparison of the voltage pairs between both sides. The resulting resistances are displayed in panel c. It is apparent that the superconducting transition does not occur at two district steps defined by T_c^a and T_c^c, but instead decreases monotonically as the current path is directed towards the a axis. This suggests that the underlying mechanism which causes the enhancement of the superconducting transition can continuously tune T_c and is not restricted to only the c axis.

As discussed above in Sect. 4.1.4, one of the hypotheses that has been made in the literature for the origin of the higher temperature resistive transition in $CeIrIn_5$ is filamentary or surface superconductivity. At the end of this chapter (c.f. Fig. 4.30),

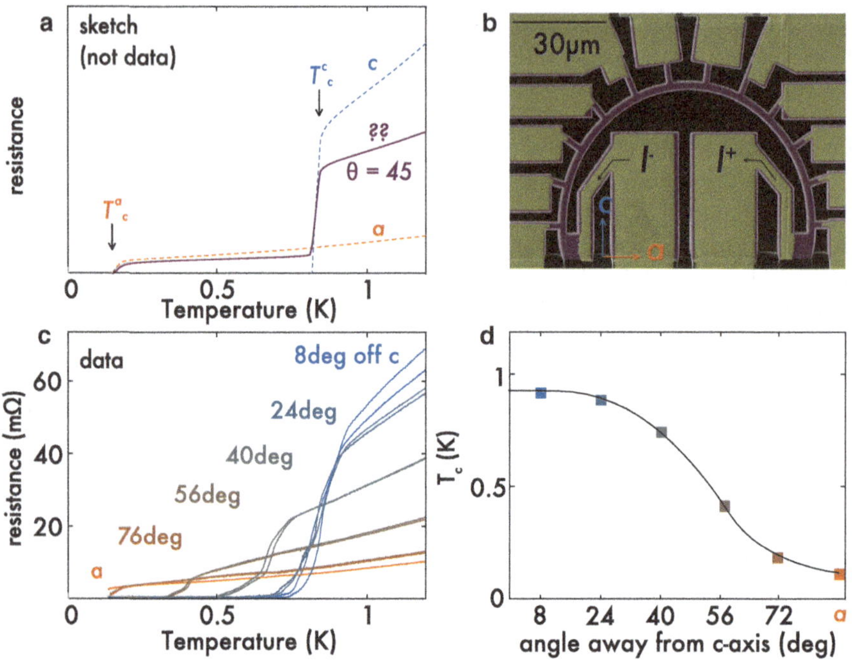

Fig. 4.12 **a** Sketch of the speculative superconducting transition along the $\theta = 45$ degree direction, which could show two distinct transitions at T_c^c and T_c^a. **b** A semi-circular transport bar was cut into an *a*-*c* CeIrIn$_5$ lamella. **c** and **d** the temperature dependent voltage drop between adjacent contacts reveals a monotonic decrease of the superconducting transition temperature as the current is directed from parallel to the *c* axis toward the *a* axis

however, I will show that the observed anomaly in macroscopic samples is in fact the result of surface damage during the fabrication process. To examine whether similar effects may need to be considered for microstructured samples, I will report on measurements of the superconductivity in microstructured CeIrIn$_5$ devices and probe their robustness against high current densities and in a magnetic field in the next sections. Due to the small volume fraction, superconducting surfaces or filaments can typically carry less supercurrent and have a dissimilar magnetic field dependence compared to the bulk superconducting material.

4.2.3 *Critical Current*

Critical current measurements of a microstructured CeIrIn$_5$ device are presented in Fig. 4.13. Remarkably, the superconducting state in CeIrIn$_5$ can sustain comparatively high currents before the zero-resistance state is destroyed. To minimise self-heating of the sample, rectangular current pulses of 83 μs were applied, fol-

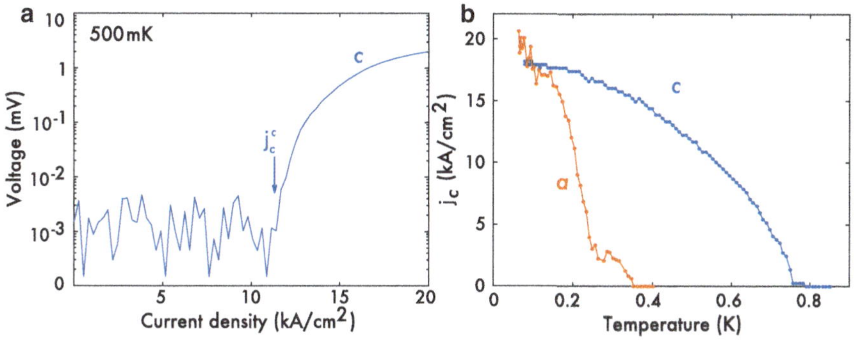

Fig. 4.13 **a** An I-V measurement reveals the critical current density j_c necessary to destroy the superconducting state. **b** The temperature dependent critical current along the a and c axes of a 'U-structure' show a similar superconducting anisotropy as the critical temperatures T_c^c

lowed by a 100 ms cool down time between pulses. Panel a shows a typical current-voltage ($I - V$) characteristic of a bar aligned along the c direction at 500 mK. The current was increased until an observable voltage signalled the breakdown of the zero-resistance state. A robust superconducting state is detected up to a critical current density of $j_c^c \approx 12.5\,\text{kA/cm}^2$. Upon cooling, c.f. panel b, j_c increases up to 18–20 kA/cm^2. Because the critical current decreases monotonically with increasing temperature, the obtained values represent a lower bound of their magnitude in the absence of heating. High critical current densities are common for heavy fermion superconductors as seen for instance in UPt_3 ($\approx$3.8 kA/cm^2), in URu_2Si_2 ($\approx$24 kA/cm^2) [51], or in $CeCu_2Si_2$ ($\approx$1–5 kA/cm^2) [52]. In conclusion, the observation of robust, high j_c superconductivity, supports a scenario of a bulk-like superconductivity over the formation of sparse superconducting filaments or surface superconductivity induced by FIB fabrication.

4.2.4 Superconducting Transition in a Magnetic Field

The upper critical field of macroscopic $CeIrIn_5$ samples has been characterised by several experimental probes including electrical resistivity, ac magnetic susceptibility χ_{ac} and specific heat C measurements [21]. Although the values of T_c and H_{c2} extracted from the resistivity are notably higher than the values extracted from thermodynamic measurements, the respective anisotropies between the $H||a$ and $H||c$ directions are remarkably similar. As it is unlikely that a secondary phase embedded in the crystal exhibits the same anisotropy as the bulk, this suggests that both transitions arise from a common underlying electronic structure. A similar observation was made by Shishido et al. [53], who suggested that the enhanced T_c may be the result of a compressed surface layer.

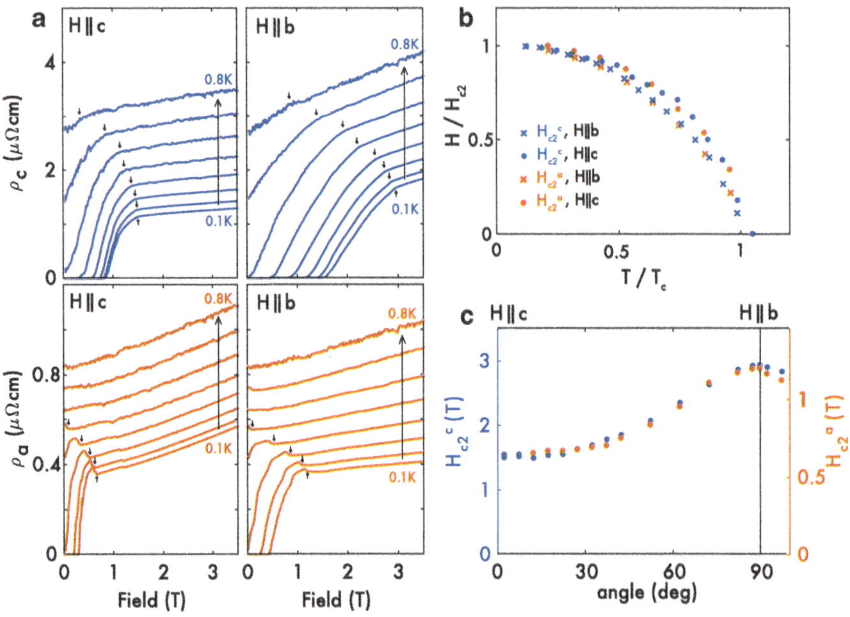

Fig. 4.14 **a** The suppression of the superconducting transition in ρ_a and ρ_c with applied magnetic field along $H||c$ and $H||b$ respectively is shown for temperatures between 0.1 K and 0.8 K in steps of 100 mK. b denotes the second in-plane axis, along which the current is not flowing. The arrows indicate the upper critical field H_{c2} values which were extracted for panel b. **b** The normalised temperature-field evolution of the superconducting transition reveals that the temperature dependence of the a and c axis critical fields is similar for both the $H||b$ and $H||c$ orientation. **c** The angle dependence of the upper critical field follows an anisotropic effective mass model with $m_c^*/m_a^* \approx 3.8$

Moreover, Shishido et al. reported the full angular dependence of H_{c2} and found that it could be well explained by an effective mass model [54] for an ellipsoidal Fermi surface, similar to what was found for $CeCoIn_5$ [55]. Using the expression

$$H_{c2}(\theta) = \frac{H_{c2}(\theta = 90^\circ)}{\sqrt{\sin^2\theta + \frac{m_c^*}{m_a^*}\cos^2\theta}}, \tag{4.12}$$

where $\theta = 90^\circ$ corresponds to $H||a$, an effective mass ratio of $\frac{m_c^*}{m_a^*} = 4.8$ can be extracted.

A similar picture unfolds for microstructured $CeIrIn_5$ devices. The resistivity as a function of magnetic field for temperatures between 100 mK and 800 mK is presented in Fig. 4.14a. The arrows indicate the position at which the values of H_{c2} were extracted. Panel b displays the normalised field-temperature evolution of the superconducting transition. Evidently, the behaviour of H_{c2} with the field applied along $H||b$ (crosses) or $H||c$ (circles) is independent of the direction in which the current is flowing. As for the macroscopic samples, this suggests that the superconducting transition along the a and c axis is caused by the intrinsic electronic structure. This is further corroborated by the nearly identical angle dependence of the upper critical field, shown in panel c. The absolute values can be simply scaled by a factor 2.5 to lie on top of each other. A fit to the data according to Eq. (4.12) yields $\frac{m_c^*}{m_a^*} = 3.8$ in both cases, similar to the previously reported value.

With a strong magnetic field and at dilution refrigerator temperatures, we can gain further insight into the electronic properties of microstructured $CeIrIn_5$ devices by investigating their Shubnikov-de Haas oscillations. These are presented in the next section, following a brief introduction to quantum oscillations.

4.2.5 Quantum Oscillations

Fundamentally, quantum oscillations arise due to the quantisation of quasiparticle cyclotron motion in a magnetic field B [56]. As a result, the quasiparticles occupy highly degenerate energy states, so-called Landau levels. This in turn causes the density of states at the Fermi energy to vary periodically in $\frac{1}{B}$, as an increasing magnetic field decreases the number of occupied Landau levels. Onsager [57] showed that, in the semiclassical limit that applies to the measurements I present, the quantum oscillation frequency F is related to the extremal cross section of the Fermi surface A_k in the plane perpendicular to the applied magnetic field

$$F = \frac{\hbar A_k}{2\pi e}. \tag{4.13}$$

This connection has made the study of quantum oscillation frequencies an extremely powerful tool for investigating Fermi surface topologies. However, for oscillations to occur in the first place, the sample must be pure enough. This is intuitively understandable, as the charge carriers have to revolve around the Fermi surface without losing phase information, i.e. without scattering, before interference can occur. This requires the phase coherent mean free path l_φ to be on the order of the circumference L_c of the cyclotron orbit. Formally, the impurity damping of the oscillation amplitude is described with the Dingle factor R_D, which can be expressed as

$$R_D = e^{-\frac{p\pi}{\omega_c \tau}}, \tag{4.14}$$

where p indicates the harmonic number. Using $\omega_c = \frac{eB}{m^*}$ and $\hbar k_F = m^* v_F$, as well as $v_F \tau \equiv l_\varphi$, expression (4.14) can be written as

$$R_D = e^{-\frac{pL_c}{2l_\varphi}}, \tag{4.15}$$

where $L_c = \frac{2\pi\hbar k_F}{eB}$ is the circumference of a circular cyclotron orbit. Since L_c is proportional to the square root of the Fermi surface area and hence the square root of the quantum oscillation frequency, $L_c \propto \sqrt{A_k} \propto \sqrt{F}$, we conclude that, for a given mean free path, high frequency oscillations are much harder to observe than low frequencies from small Fermi surfaces. Moreover, finite temperatures and finite quasiparticle lifetimes, or local variations in the chemical potential, lead to phase smearing effects. This describes the broadening of the oscillation frequency distribution, which leads to a damping and ultimately complete suppression of the oscillation amplitude.

In summary, the exponential suppression of the oscillation amplitude with decreasing mean free path imply that quantum oscillations are delicate and a stringent test for sample purity and homogeneity.

Quantum oscillations can, in principle, be detected in any property that depends on the oscillating density of states at the Fermi level. For instance, the oscillations in the magnetisation $M = -\partial F/\partial B$, called de Haas-van Alphen (dHvA) oscillations, occur due to oscillations in the free energy F of the system. Alternatively, Shubnikov-de Haas (SdH) oscillations, which are oscillations of the resistivity, have their origin in the fact that the quasiparticle scattering rate depends on the density of states at the Fermi level. Due to high conductivity and hence low resistance of macroscopic samples, these are often challenging to measure. It can happen that the currents needed to achieve the necessary signal-to-noise ratio are incompatible with dilution refrigerator measurements. Microstructured devices, however, typically have a much larger total device resistance due to the large geometrical factors achievable in microbars and thus provide ideal conditions for the observation of quantum oscillations.

In Fig. 4.15 I present Shubnikov-de Haas oscillations of microstructured CeIrIn$_5$, which are the first ones to be reported (at all) to the best of our knowledge. Below 125 mK and above 8 T the oscillations are clearly detectable for currents parallel to both the a and c directions, as indicated by the background subtracted resistivity signal shown in panel a. The fast Fourier transform reveals clear peaks in the FFT spectrum, presented in panel b, corresponding to the extremal Fermi surface orbits, according to Eq. (4.13). Clearly, large Fermi surface sections of up to 6–7 kT are observed, evidencing the high quality of the microstructure. The full angular dependence of the frequency spectrum is displayed in panel c (red and blue data points) and is compared to the previously reported dHvA frequencies (grey line and annotation) [30]. A low frequency band disperses from 270 T for $H||c$ to 460 T for $H||b$, corresponding to the γ-branch observed in dHvA, hosting the lightest quasiparticles in CeIrIn$_5$. Moreover, the large, heavy orbits and their fine structure are readily observed and are in agreement with the dHvA data. This suggests that CeIrIn$_5$ can be microstructured without deterioration of its crystal quality and that the cross sectional Fermi surface area remains unchanged.

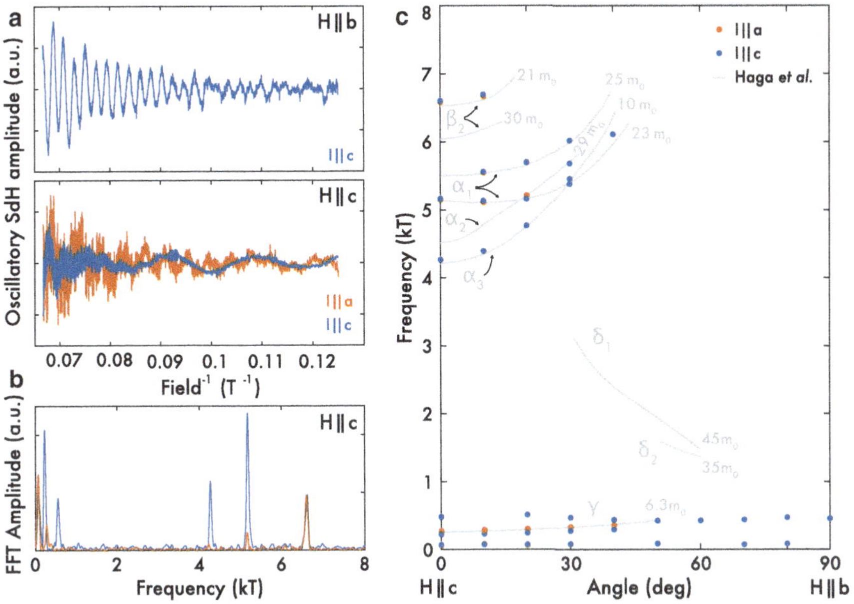

Fig. 4.15 **a** The magnetoresistance measured at 125 mK reveals Shubnikov-de Haas oscillations after a 4th order polynomial background was subtracted. **b** The fast Fourier transform shows the frequency components of the data displayed in A. **c** The angular dependence of the quantum oscillations in a microstructured sample of $CeIrIn_5$ are compared with the results of a dHvA measurement of a macroscopic sample by Haga et al. [30]. The identified Fermi surface sections α_i, β_i, γ and δ_i are labeled and their effective masses, determined by dHvA, are indicated next to them

4.2.6 *Differential Thermal Contraction Mismatch*

So far we have seen that the enhanced T_c^c transition into a zero-resistance state consistently occurs in every microstructured device and cannot be explained by systematic fabrication artefacts. The presence of superconducting filaments, originally suggested to occur in macroscopic samples [31], is at odds with the high critical current densities as well as the angle dependence of H_{c2} observed in microstructures. Moreover, the observation of SdH oscillations, which are in good agreement with dHvA oscillations measured in macroscopic crystals, clearly indicates that the high quality and stoichiometry of the microstructured $CeIrIn_5$ devices.

One aspect that we have not considered so far is that microstructures are typically in firm contact with their substrate. The thermal expansion $\Delta L/L$ between sapphire and a typical metal such as $CeIrIn_5$ can be reasonably large, as shown in Fig. 4.16a. The differential thermal contraction mismatch can lead to a significant strain at the interface of the materials, which may effect their electronic properties. In our case, $CeIrIn_5$ contracts much more than sapphire and hence the lamella will effectively experience biaxial tensile strain at low temperatures, c.f. Fig. 4.16b. This effect is well-known in epitaxial multilayers and is exploited in strained silicon technology,

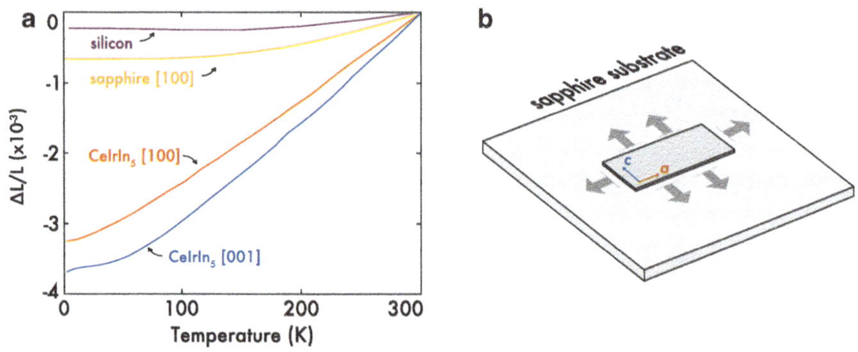

Fig. 4.16 **a** Temperature dependence of the thermal expansion of CeIrIn5 [59] and two typical substrate materials, sapphire [60] and silicon [61]. **b** As a result of differential thermal contraction between the CeIrIn$_5$ and the sapphire substrate, the lamella will experience a biaxial tensile strain at low temperatures

in which transistors based on strained Si experience a more than 100% increase of the electron mobility [58].

In the following, I will explore the possibility of biaxial strain as the source of the enhanced T_c^c in CeIrIn$_5$ lamellae by fabricating strain free devices. Note that this is consistent with the observed quantum oscillation frequency spectrum, which was shown to remain completely unchanged as a function of pressure up to the quantum critical point [62].

4.2.7 *Strain-Free Microstructures*

To investigate the role of substrate strain on a microstructured sample, I fabricated a strain-free structure. This, however, is not a straightforward problem to solve. In order to perform an electrical transport measurement at cryogenic temperatures, at least four leads must make a low resistance connection to the sample for a four-point measurement and for thermal anchoring. Even the thinnest metallic wires commonly available (10–12 μm diameter) are still a factor of 3–4 thicker than the lamella, such that undesired torque or stress effects can not be excluded.

An elegant solution is presented in Fig. 4.17a. A standard CeIrIn$_5$ lamella was cantilevered off the edge of a sapphire substrate. It is held in place by a small amount of Araldite epoxy, which additionally creates a smooth transition from the substrate to the top of the lamella. Crucially, during fabrication, the epoxy is first dried completely for 24 h before it is cured, to avoid capillary drag of the lamella onto the substrate during epoxy reflow. The sample is top-contacted using gold sputter deposition (overlaid in yellow). The final step is to pattern long thin flexures into the lamella to relieve the thermal expansion mismatch induced strain. For this device we have chosen to pattern a Montgomery square (purple) into the far end of the lamella.

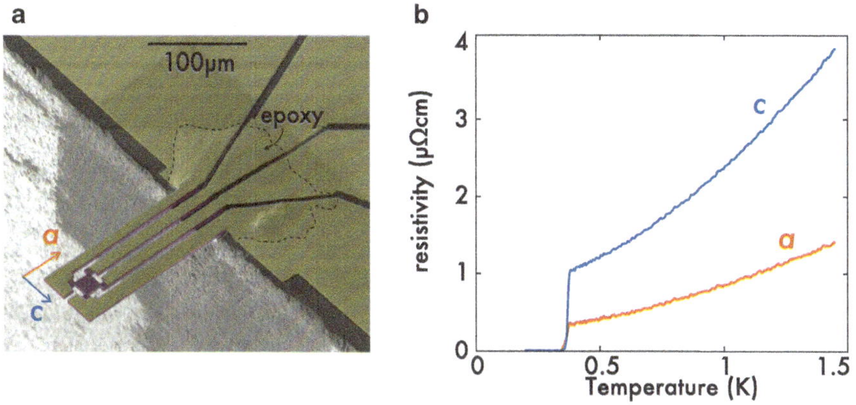

Fig. 4.17 **a** In order to completely eliminate strain from differential thermal contraction, we have fabricated a free standing device, by cantilevering a lamella off the edge of a sapphire substrate. The lamella was patterned, such that long, thin beams act as strain relief flexures and connect a Montgomery square to the contact pads. **b** The temperature dependent resistivity along the a and c axis of a strain-free $CeIrIn_5$ sample shows an isotropic superconducting transition at exactly $T_c = T_c^{bulk} = 0.38$

This permits determination of the a and c axis resistivity using only four contacts. The temperature dependent resistivity is is displayed in Fig. 4.17b. To the best of my knowledge, this is the first and so far only time when the transition into a zero-resistance state in $CeIrIn_5$ has been measured at exactly the thermodynamic value of $T_c^{bulk} = 0.38\,K$. Samples prepared in this fashion are well suited for cryogenic measurements in DC fields and, if the epoxy is replaced by platinum leads, could potentially also work under hydrostatic pressure environments. In pulsed magnetic fields, on the other hand, the vibrations and experienced torque of the sample during the pulse will presumably induce a significant amount of noise in the voltage channel, rendering resistivity measurement challenging at best.

This measurement clearly demonstrates the significance of substrate strain to observed shifts in T_c in microstructured samples. As a bulk probe, however, resistivity measurements are not sensitive to the underlying details of the strain distribution, and hence we have collaborated with the groups of K. Nowack and B. J. Ramshaw at Cornell University, who have performed finite element simulations and scanning SQUID microscopy measurements, which present a detailed picture of the underlying strain variation in microstructured lamellae. Their results will be presented in Sects. 4.3 and 4.4.

4.2.8 Microstructures of Related '115' Compounds

We have seen that $CeIrIn_5$ is sensitive to anisotropic strain introduced in microstructures by a thermal expansion mismatch. Based on the highest measured $T_c^c \approx 1.1$ K (c.f Fig. 4.9a) in a microstructure and the uniaxial pressure study discussed in Fig. 4.8a, we can infer that a micron sized device can locally experience up to an equivalent of 10 kbar of local pressure. A natural concern is then the strain sensitivity of $CeRhIn_5$ and $CeCoIn_5$ in microstructured devices. The low-temperature resistivity data of these compounds measured in microstructured samples were kindly provided by P. Moll ($CeRhIn_5$) and C. Putzke ($CeCoIn_5$) and are displayed in Fig. 4.18a and b.

The microstructured $CeRhIn_5$ device shows no superconducting transition down to 100 mK, in Fig. 4.18a. This is in line with the measurements of Knebel et al. [63], which show that over 8 kbar of hydrostatic pressure are necessary to induce superconductivity. This, however, challenges the findings of Paglione et al. [64], which report ambient pressure superconductivity below 110 mK. Wire saw induced internal strain would offer a natural explanation for their results.

Microstructured $CeCoIn_5$, in Fig. 4.18b, does not show an anisotropic T_c behaviour. The reason for this is that the uniaxial pressure dependence of T_c in the Co compound is smaller and positive regardless of the strain direction, c.f. Table 4.1. This means that strain will affect the superconducting transition less and can only cause it to increase, but not decrease, in contrast to the Ir compound. Additionally, it was shown [65] that the uniaxial pressure dependence is non-linear in $CeCoIn_5$, such that the value of T_c is roughly the same at 0 and at 4 kbar. This further decreases the effect of local anisotropic strain on the transition temperature.

In summary, we have seen that the Ce-based 115 family is sensitive to strain, such that great care should be taken in the preparation of these samples. In order to investigate the role of substrate strain on lamellae and hence on the value of T_c of $CeIrIn_5$, we

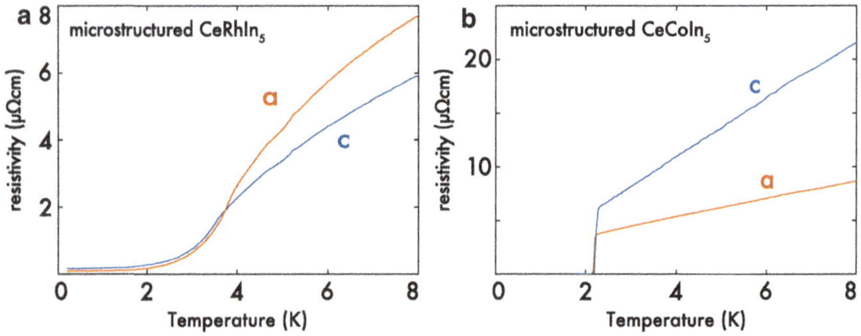

Fig. 4.18 Low temperature resistivity data of microstructured (**a**) $CeRhIn_5$ and (**b**) $CeCoIn_5$ lamellae, with similar dimensions and fabricated identically as the devices in Fig. 4.9. The data presented in these figures were measured by P. Moll and C. Putzke

Table 4.1 Comparison of the uniaxial pressure dependence of the superconducting transition temperatures in $CeIrIn_5$ and $CeCoIn_5$. * denotes the non-linear behaviour of T_c with respect to uniaxial strain

	T_c (mK)	$\partial T_c/\partial p_a$ $(\frac{mK}{kbar})$	$\partial T_c/\partial p_c$ $(\frac{mK}{kbar})$	$\partial T_c/\partial p_{hydro}$ $(\frac{mK}{kbar})$	references
$CeIrIn_5$	380	+56	−66	+25	[42, 66]
$CeCoIn_5$	2230	+29	+7.5*	+50	[41, 65, 67]

have collaborated with the scanning SQUID microscopy group led by K. Nowack at Cornell University. With the ability to locally measure Meissner screening, they can create a microscopic T_c map of a $CeIrIn_5$ lamella. Additionally, a local strain map can be calculated using a finite element software package. The results can be compared by converting the local strain into a simulated T_c map, by using the results of Dix et al. [42], who measured the superconducting transition as a function of uniaxial strain.

In the following I will first discuss the implementation and results of the finite element simulations to gain insight and an understanding of the local strain of a biaxially tensioned lamella. I will then compare the simulated results with the measured T_c maps acquired by scanning SQUID microscopy.

4.3 Finite Element Simulations

The following section is based on the work of F. Theuss at Cornell University, whose finite element calculations form the basis of the results presented here. Moreover S. Ghosh has determined the elastic moduli of $CeIrIn_5$ by conducting a resonant ultrasound experiment under the supervision of B. J. Ramshaw at Cornell University.

The aim of the calculation is to obtain a map of the local superconducting transition temperature $T_c(\vec{r})$ of a $CeIrIn_5$ lamella, which is epoxy bonded to a sapphire substrate and cooled down to cryogenic temperatures. The differential thermal contraction mismatch between the substrate and the sample will generate a biaxial strain, which in turn will cause local variations of T_c across the lamella. In a first step, the strain field $\varepsilon(\vec{r})$ in the lamella is calculated using finite element simulations with realistic input parameters for the elastic moduli C_{ij} and the thermal expansion coefficients $\alpha_i(T)$. In a second step, the strain field $\varepsilon(\vec{r})$ can directly be converted into the local c/a ratio, which can then be transformed into a T_c map, based on the measurements of Dix et al. [42] (c.f. Fig. 4.22). This procedure is graphically summarised in Fig. 4.19.

Eventually the calculated dependence of the superconducting transition will be compared to the measured T_c map determined by local susceptibility measurements (c.f. Sect. 4.4). If the calculations are indeed validated by the experiment, this then permits further numerical investigation of the influence of experimental parameters, such as epoxy properties (e.g. thickness, stiffness) or alternative substrate materials.

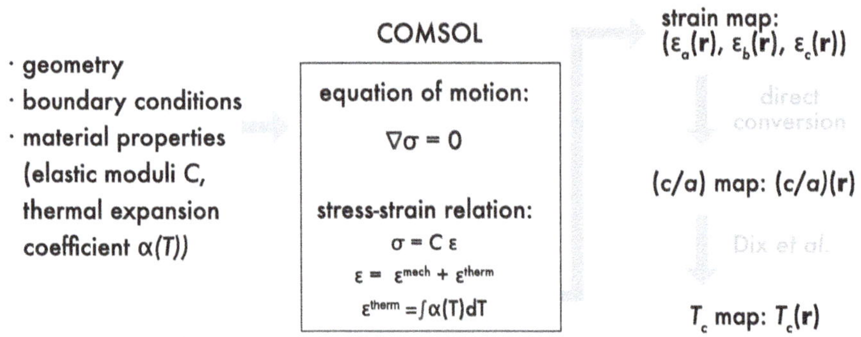

Fig. 4.19 Graphical summary of the calculation strategy. With appropriate material properties and boundary conditions, finite element software will determine the strain field $\varepsilon(\vec{r})$ of a system. This can then be converted into a local map of T_c if the uniaxial pressure dependence of the material under investigation is known

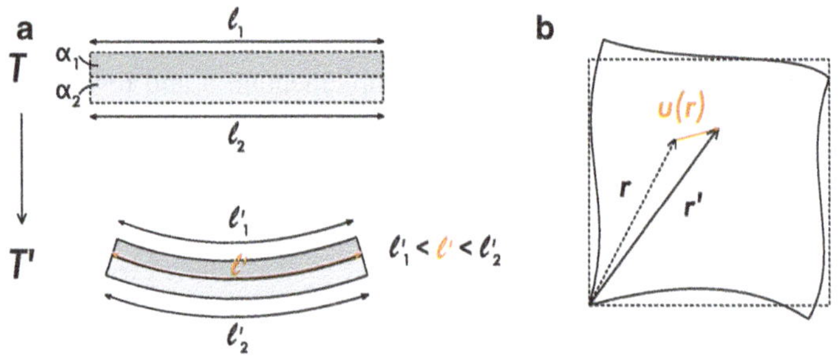

Fig. 4.20 **a** Two connected materials with different thermal expansion coefficients α_1 and α_2 will experience a misfit strain at their interface, when subject to a temperature change. **b** In general the deformation of an object can be described via a displacement field $u(\vec{r})$

In particular, numerical simulations enable an examination of tailored strain in diverse device geometries, which can guide the fabrication of strain modulated structures.

The remainder of this section will begin with a brief introduction to strain and stress in elastic media. Following this, I will give an overview of the thermoelastic strain simulations performed using the finite element analysis software, COMSOL, and show how the local strain tensor can be converted into a local T_c map. Finally, the results of the thermoelastic simulations on CeIrIn$_5$ are presented at the end of this section.

4.3.1 Strain and Stress in Elastic Media

When two materials with distinct thermal expansion coefficients α_1 and α_2 are bonded together and then subject to a temperature change, they will in general expand or contract by different amounts. This so-called differential thermal contraction (or expansion) leads to a strain at the interface of the materials. This is a commonly known effect exploited for instance in bimetallic thermostats, sketched in Fig. 4.20a. In this one-dimensional model example, a change in temperature $\Delta T = T' - T$ causes the top and bottom layer to contract by different amounts and cause the system to be strained internally. Whilst the individual components would freely contract by a length $\ell_i - \ell_i' = \Delta\ell_i = \alpha_i \Delta T$ in linear response, the composite system experiences a strain, which is strongest at the interface. The misfit strain can be estimated by $\varepsilon = (\alpha_1 - \alpha_2)\Delta T$. In two dimensions, biaxial strain effects are commonly studied in epitaxially strained thin films [68–70] and exfoliated 2D materials [71, 72] but rarely in bulk materials [73].

Regardless of the dimensionality, any deformation of an object can be described in terms of a displacement field $\vec{u}(\vec{r}) = \vec{r}' - \vec{r}$, which is a three component vector expressing the local displacement as the difference of the position vectors $\vec{r}$ and $\vec{r}'$, c.f. Fig. 4.20b. By decomposing the displacement field into its three spatial coordinates $i = x, y, z$, it follows that

$$u_i(r) = r_i' - r_i. \tag{4.16}$$

In general, within a strained solid, the distance between two coordinate points is changed, i.e. if two closely spaced points were separated by dr_i, their distance will be modified to $dr_i' = dr_i + du_i$ following the deformation. If the distance between two points prior to deformation is defined as $dl = \sqrt{dr_1^2 + dr_2^2 + dr_3^2}$, then the distance dl' after the distortion can be calculated

$$\begin{aligned} dl'^2 &= \sum_i dr_i'^2 = \sum_i (dr_i + du_i)^2 \\ &= dl^2 + 2\sum_{i,j} \frac{\partial u_i}{\partial r_j} dr_i dr_j + \underbrace{\sum_{i,j,k} \frac{\partial u_i}{\partial r_j}\frac{\partial u_i}{\partial r_k} dr_j dr_k}_{\text{neglect for small deformations}} \end{aligned}$$

$$\Rightarrow dl'^2 - dl^2 = 2\varepsilon_{ij} dr_i dr_j, \qquad \varepsilon_{ij} \equiv \frac{1}{2}\left(\frac{\partial u_i}{\partial r_j} + \frac{\partial u_j}{\partial r_i}\right), \tag{4.17}$$

where the last step makes use of the fact that indices in the double sum $\sum_{ij}$ can be exchanged. The second rank tensor ε_{ij} introduced in (4.17) is called the strain tensor. Because it is symmetric, $\varepsilon_{ij} = \varepsilon_{ji}$, the strain tensor is locally diagonalisable at every point (but not necessarily at all points simultaneously). In other words, at any given point the coordinate system can be chosen in a way such that the off-diagonal components are zero and only the diagonal elements ε_{ii} remain. Around such a point Eq. (4.17) simplifies to

$$dl'^2 = (1 + 2\varepsilon_{xx})dr_x^2 + (1 + 2\varepsilon_{yy})dr_y^2 + (1 + 2\varepsilon_{zz})dr_z^2.$$

It is then straightforward to decompose this into the individual coordinate components and find the relative change along the ith axis

$$\begin{aligned} dr_i' &= \sqrt{1 + 2\varepsilon_{ii}}\, dr_i \\ \Rightarrow \frac{dr_i'}{dr_i} &= \sqrt{1 + 2\varepsilon_{ii}} \quad \approx 1 + \varepsilon_{ii}, \end{aligned} \tag{4.18}$$

where in the last step quadratic terms in ε can be omitted for small deformations. We will later make use of this expression, when converting the calculated strain field $\varepsilon(\vec{r})$ into the local c/a ratio.

The stress tensor $\sigma(\vec{r})$ describes how the forces F acting on a system are internally distributed. Individual elements σ_{ij} express the stress acting along the direction j on the plane normal to i. Hence the diagonal components σ_{ii} describe normal stresses, whereas off-diagonal components are responsible for shear stress. In the case where differential thermal contraction is solely responsible for the strain in the system and there are no external forces applied, the total force F integrated over the volume of a system must equate to zero. Within the theory of linear elasticity the equation of motion which we want to solve is then given by

$$\begin{aligned} \nabla \cdot \sigma + \underbrace{F}_{=0} &= \rho \underbrace{\ddot{u}(\vec{r})}_{=0} \\ \Rightarrow \nabla \cdot \sigma &= 0. \end{aligned} \tag{4.19}$$

As we are dealing with a static problem, we can set $\ddot{u}(\vec{r}) = 0$. The strain and stress tensors of a system are related by the material-dependent rank four elasticity tensor C,

$$\sigma_{ij} = C_{ijkl}\, \varepsilon_{kl}, \tag{4.20}$$

where the components C_{ijkl} are called the elastic moduli. The number of independent elastic moduli is reduced by the underlying symmetry of the crystal lattice, such that in a tetragonal system, for instance, Eq. (4.20) can be expressed by only six independent parameters. In matrix form Eq. (4.20) reads

$$\begin{pmatrix} \sigma_{xx} \\ \sigma_{yy} \\ \sigma_{zz} \\ \sigma_{yz} \\ \sigma_{zx} \\ \sigma_{xy} \end{pmatrix} = \begin{pmatrix} C_{11} & C_{12} & C_{13} & 0 & 0 & 0 \\ C_{12} & C_{11} & C_{13} & 0 & 0 & 0 \\ C_{13} & C_{13} & C_{33} & 0 & 0 & 0 \\ 0 & 0 & 0 & C_{44} & 0 & 0 \\ 0 & 0 & 0 & 0 & C_{44} & 0 \\ 0 & 0 & 0 & 0 & 0 & C_{66} \end{pmatrix} \begin{pmatrix} \varepsilon_{xx} \\ \varepsilon_{yy} \\ \varepsilon_{zz} \\ \varepsilon_{yz} \\ \varepsilon_{zx} \\ \varepsilon_{xy} \end{pmatrix}. \tag{4.21}$$

Table 4.2 Definition of the contracted Voigt notation for labelling the components of a symmetric tensor

Tensor	11	22	33	23, 32	13, 31	66
Voigt	1	2	3	4	5	6

Here, I have used the contracted Voigt notation, summarised in Table 4.2, which simplifies the labelling of the tensor components C_{ijkl} by reducing the number of indices. The elastic properties of a tetragonal material, such as $CeIrIn_5$ are therefore fully described by the normal modes C_{11}, C_{33}, C_{44}, C_{66} and the shear modes C_{12} and C_{13}. These can be determined for instance in pulsed echo or resonant ultrasound experiments.

The last element we need is to include into our discussion is the thermal expansion of the individual subsystems. This is achieved by adding the strain due to thermal expansion $\varepsilon^{\text{therm}}$ to the total strain of the system

$$\varepsilon = \varepsilon^{\text{mech}} + \varepsilon^{\text{therm}}, \tag{4.22}$$

where the thermal strain of a subsystem is determined by its temperature-dependent thermal expansion coefficient $\alpha(T)$

$$\varepsilon^{\text{therm}} = \int_{T_0}^{T\prime} \alpha(T)dT. \tag{4.23}$$

With this the constitutive Eq. (4.20) is then given by

$$\sigma = C\left(\varepsilon - \int_{T_0}^{T\prime} \alpha(T)dT\right)$$

and we are now fully set to calculate the strain field in a $CeIrIn_5$ lamella induced by thermal contraction mismatch to the sapphire substrate.

4.3.2 *Thermoelastic Strain Simulation Using COMSOL*

The following section is intended to give detailed information on the modelling parameters and settings used within the structural mechanics module of the commercially available finite element software COMSOL. A reader purely interested in the outcome of the calculations may wish to skip forward to Sect. 4.3.4, where the results of the simulation are presented.

The elastic moduli of $CeIrIn_5$ were determined in a resonant ultrasound experiment in which the resonance frequencies of the crystal are measured. Each frequency is uniquely determined by the sample dimensions, density, and elastic moduli, and can

Table 4.3 The six elastic moduli of CeIrIn$_5$ at 295 K were determined by S. Ghosh and B. J. Ramshaw through a resonant ultrasound (RUS) measurement

Elastic modulus	C_{11}	C_{33}	C_{12}	C_{13}	C_{44}	C_{66}
Value (GPa)	153.1	138.8	41.3	53.1	46.2	42.0

Table 4.4 Summary of the anisotropic elastic properties and density of CeIrIn$_5$, sapphire and epoxy resin. For simplicity, the mechanical properties of sapphire are approximated as isotropic

	CeIrIn$_5$	Sapphire (Al_2O_3) [61]	Epoxy [78]	Units
Young's modulus E	$E_a = 129$; $E_c = 109$	470	7.2	GPa
Poisson's ratio ν	$\nu_{aa} = 0.16$; $\nu_{ac} = 0.32$; $\nu_{ca} = 0.27$	0.24	0.33	–
Shear modulus G	$G_{ac} = 46.2$	147.4	3.7	GPa
Density ρ	2800	3970	1800	kg/m^3

be calculated by solving the 3D elastic equation [74]. Conversely, the elastic moduli can be determined from the resonances by solving the inverse problem following the procedure described in [75]. The results of the measurement are summarised in Table 4.3.

Given the six independent elastic moduli of a tetragonal crystal, the six independent components of Young's modulus E, the shear modulus G and Poisson's ratio ν (or an equivalent set of moduli) can directly be determined. In a tetragonal system with in-plane lattice direction a and out-of-plane direction c, the relation between the elasticity tensor C (defined in Eq. (4.21)) and the moduli E_i, ν_{ij}, G_{ij} is given by [76, 77]

$$C^{-1} = \begin{pmatrix} 1/E_a & -\nu_{aa}/E_a & -\nu_{ca}/E_c & 0 & 0 & 0 \\ -\nu_{aa}/E_a & 1/E_a & -\nu_{ca}/E_c & 0 & 0 & \\ -\nu_{ac}/E_a & -\nu_{ac}/E_a & 1/E_c & 0 & 0 & 0 \\ 0 & 0 & 0 & 1/G_{ac} & 0 & 0 \\ 0 & 0 & 0 & 0 & 1/G_{ac} & 0 \\ 0 & 0 & 0 & 0 & 0 & 2(1+\nu_{aa})/E_a \end{pmatrix} . \quad (4.24)$$

With this the COMSOL input parameters (E,ν,G) can directly be determined and are listed in Table 4.4 alongside the elastic properties of sapphire and epoxy resin. For simplicity, we approximate the sapphire as mechanically isotropic. As the normal of the sapphire substrate is oriented along the (0001) direction, this approximation is justified since we are mainly concerned about the in-plane, bi-axial tension exerted on the lamella and neglect the anisotropic contraction of the sapphire along the c direction.

The remaining input parameters needed for the simulation are the thermal contraction coefficients $\alpha(T)$ of the constituent materials. For CeIrIn$_5$ and sapphire, the

Table 4.5 Summary of the total thermal contraction of $CeIrIn_5$, sapphire and epoxy resin between room temperature and 4 K

	Total thermal contraction $(\Delta L/L_{293\,K})$
$CeIrIn_5$	$\varepsilon_0^a = -3.2\times10^{-3}$, $\varepsilon_0^c = -3.5\times10^{-3}$ [59]
Sapphire	$\varepsilon_0 = -8\times10^{-4}$ [60]
Epoxy	$\varepsilon_0 = -1.16\times10^{-2}$ [61]

temperature-dependent thermal contraction $\Delta L(T_0 - T')/L_{T_0} = \int_{T_0}^{T'} \alpha(T)dT$ is displayed in Fig. 4.16. The temperature-dependent values for a typical epoxy resin can be found in [61]. For a direct comparison the values of the total thermal contraction ε_0 between between room temperature and 4 K are summarised in Table 4.5. The contraction of sapphire is approximately an order of magnitude less than that of $CeIrIn_5$, which in turn contracts much less than the epoxy. The epoxy, however, is much softer (i.e. has a smaller Young's modulus) than $CeIrIn_5$ and sapphire, such that the strain generated by contracting epoxy can essentially be neglected.

The COMSOL simulations are set up such that the composite $CeIrIn_5$/epoxy/sapphire system is unstrained at room temperature, i.e. a zero-stress boundary condition at 293 K is enforced. The boundary between the materials is modelled by requiring the strain field to be continuous at each interface. Further, we have investigated the role of the epoxy shape and height by studying two distinct geometries, displayed in Fig. 4.21. The first model shown in Fig. 4.21a omits the epoxy at the side of the lamella and only considers a thin layer ($t \sim 200$ nm), which bonds the lamella to the substrate. In contrast, the edges of the lamella shown in Fig. 4.21b are fully embedded in epoxy. We find that increasing the thickness t of the epoxy gradually attenuates the strain field in the microstructure, but does not alter the spatial strain distribution. Adding epoxy to the sides of the microstructured devices slightly modifies the strain field along the edges but does not affect its overall qualitative shape. Moreover, as the exact mechanical properties of the specific epoxy (Araldite 5 min rapid) used in this experiment have not been experimentally determined, we have instead used a range of values for its Young's modulus E in the simulation. We found, that by varying E between 1.6 and 9.2 GPa the predicted values of T_c vary less than 10 mK, which is not surprising, as the total thermal contraction of the epoxy is much smaller than that of the sapphire or the lamella. In summary, because the strain in the lamella predominantly stems from the thermal mismatch to the sapphire substrate, the precise geometry and properties of the epoxy play a minor role in the qualitative form of the strain modulation in the lamella. The main role of the epoxy is to act as soft film providing a spacer layer to relax some of the internal strain.

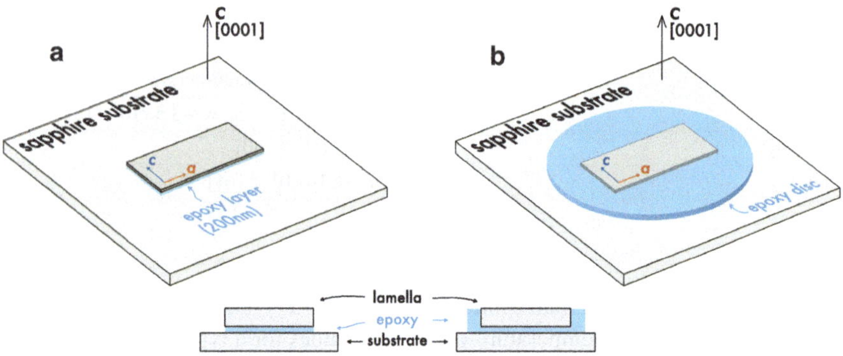

Fig. 4.21 The role of epoxy was numerically investigated in two different geometries. **a** The microstructure is held in place by a 200 nm thin layer of epoxy underneath the lamella. **b** The microstructure is embedded in a disc of epoxy, such that the strain field along the edges is modified

4.3.3 Strain to T_c Map Conversion

Given the three-component strain field $\varepsilon(\vec{r})$ of the lamella, it is then straightforward to calculate the local c/a ratio by applying the result of Eq. (4.18). The strained (c/a)' ratio can then be simply approximated in terms of the unstrained (c/a) ratio by

$$\left(\frac{c}{a}\right)' \approx \frac{1+\varepsilon_c}{1+\varepsilon_a}\left(\frac{c}{a}\right). \tag{4.25}$$

Finally, the local value of T_c can be determined by using the empirical formula

$$T_c\left(\frac{c}{a}\right) = 35\,\mathrm{K}\cdot\left(\frac{c}{a}\right) - 55.98\,\mathrm{K}, \tag{4.26}$$

based on combining (4.25) with the analysis of Dix et al. [42] presented in Fig. 4.22. Note that when a biaxial pressure is applied in the plane of $CeIrIn_5$, the basal plane undergoes a tetragonal to orthorhombic distortion and the two principal in-plane directions, a and b, are no longer equivalent.

4.3.4 Results from Finite Element Simulations

During our investigations we have numerically studied, and subsequently performed scanning SQUID microscopy measurements (c.f. Sect. 4.4), on three microstructured $CeIrIn_5$ devices. In the following I will present and discuss the result of the strain calculations performed using COMSOL. The corresponding T_c maps are displayed

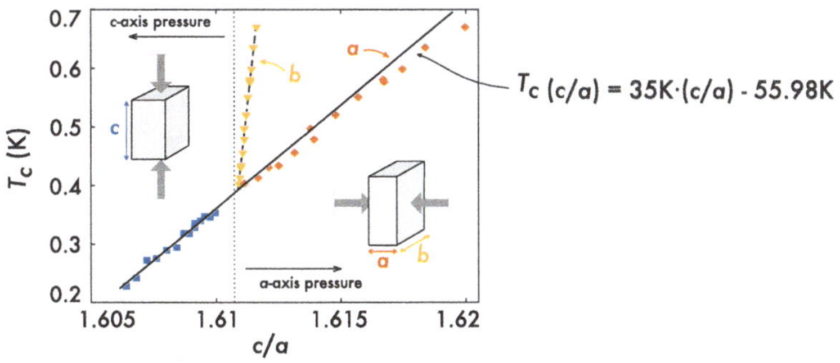

Fig. 4.22 The dependence of T_c as a function of the c/a ratio was determined by Dix et al. [42]. Note that by applying uniaxial pressure in the in-plane direction, the tetragonal symmetry of the crystal is reduced to an orthorhombic symmetry, where we denote shorter axis as a and the longer axis as b. Figure reproduced from [42]

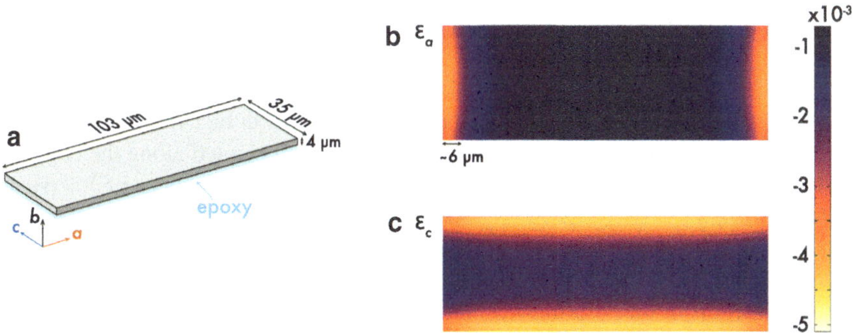

Fig. 4.23 **a** The lamella is cut along the three principal directions of the crystal and anchored to a sapphire substrate (not shown) by a 200 nm layer of epoxy. **b** and **c** show the diagonal components of the simulated strain fields ε_a, and ε_c respectively on the top surface of an unstructured $CeIrIn_5$ lamella at 0.35 K. The simulation results presented in this figure are based on work performed by F. Theuss

alongside the scanning SQUID data in the next section. The three devices are (1) an unstructured lamella (c.f. Fig. 4.23), which serves as an instructive model case from which one can learn how the strain is distributed in a comparatively simple device without further complications. Remarkably, even in this basic geometry we find that the superconducting transition temperature is modulated in a complex manner across the lamella. This sets the basis for the remaining structured devices, which are (2) a lamella patterned into a Montgomery geometry (c.f. Fig. 4.24) and (3) a modified Montgomery device with contacts in the middle, rather than at the corners of the central square (c.f. Fig. 4.25). Note that in the following I will only be presenting the top surface of the devices as these will later be investigated by scanning SQUID microscopy.

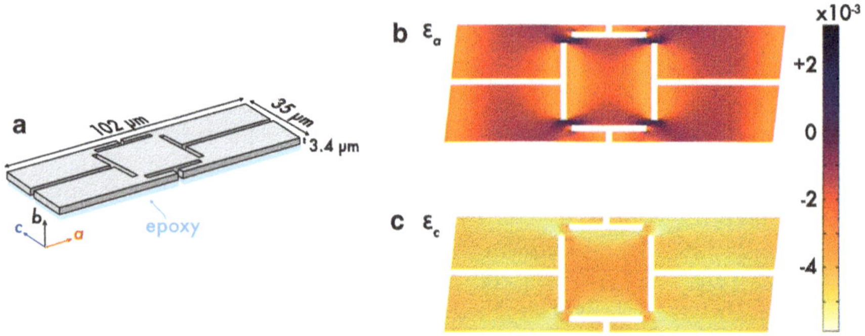

Fig. 4.24 **a** A lamella is cut such that the edges align roughly with the three principal directions of the crystal. Importantly, the epoxy layer underneath the lamella does not fill the channels that were milled with the FIB. **b** and **c** show the diagonal components of the simulated strain fields, ε_a and ε_c respectively, at the top surface of a CeIrIn$_5$ lamella in a Montgomery geometry, which is anchored to a sapphire substrate by a 200 nm layer of epoxy, at 0.35 K. The simulation results presented in this figure are based on work performed by F. Theuss

Examining the strain field in a plain lamella displayed in Fig. 4.23, we recognise that the largest strain in the a-c plane of the lamella is generated along the edges of the device and relaxes towards the centre over a characteristic distance of $\delta \sim 6\,\mu$m.

Recalling that tensile strain along the c direction increases T_c and conversely T_c is decreased when tensile strain is applied along the crystallographic a direction, the strain patterns in Fig. 4.23b and d straight away elucidate the transport results from the last section. In this example, a current running parallel to the c axis can flow along a high T_c region, while a current sourced parallel to the a axis can only flow as supercurrent at lower temperatures (c.f. Fig. 4.27 in the next section).

Our understanding of the local strain pattern is further expanded by examining the strain simulations of the Montgomery device, shown in Fig. 4.24a. As before, the bright yellow areas indicate the regions of highest tensile strain. Unlike in the plain lamella, the cuts in the Montgomery device, which define the structure, lead to regions of overall compressive strain along the a axis at the corners of the central square (dark regions in Fig. 4.24b). In general, the cuts have a large effect on the local strain pattern. In the immediate vicinity of a cut, including the outer edges of the device, the strain values are extremal and relax towards the centre. For the Montgomery device presented here, this results in spatially separated paths of high and low T_c, leading to the observed anisotropic superconducting behaviour. Hence, in principle, well-defined cuts could be used to tailor the local strain in a lamella such that complex strain landscapes can be fabricated within one lamella.

Another example of a complex strain pattern in a structured lamella is presented in Fig. 4.25. In this geometry the contacts to the central square connect to the middle rather than to the corners, which greatly influences the local strain pattern of the device. In particular, the areas where the outer contacts connect to the central square experience a strong tensile strain along one direction and simultaneously a

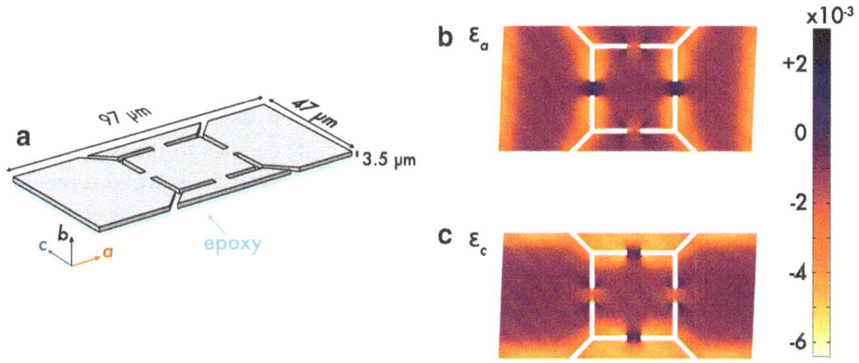

Fig. 4.25 **a** A lamella is cut such that the edges align roughly with the three principal directions of the crystal. Importantly, the epoxy layer underneath the lamella does not fill the channels that were milled with the FIB. **b** and **c** show the three diagonal components of the simulated strain fields, ε_a and ε_c respectively, at the top surface of a $CeIrIn_5$ lamella in a modified Montgomery geometry, which is anchored to a sapphire substrate by a 200 nm layer of epoxy, at 0.35 K. The simulation results presented in this figure are based on work performed by F. Theuss

compressive strain along the orthogonal direction. This demonstrates the versatility of the induced strain and hence T_c patterns, which develop at low temperatures due to differential thermal contraction.

This concludes the discussion of thermo-elastically induced strain in $CeIrIn_5$ lamellae. The next section will directly compare the simulated distribution of T_c with measured maps of the local susceptibility.

4.4 Scanning SQUID Microscopy Investigations

This section will begin with a brief introduction to scanning SQUID microscopy and subsequently the scanning SQUID measurements of the local susceptibility of the three microstructures presented in the last section will be discussed. The devices were designed and fabricated by me, and the scanning SQUID measurements and data analysis were performed by G. Ferguson in the group of K. Nowack at Cornell University.

The operating principle of a scanning SQUID microscope relies on closely rastering a highly sensitive magnetometer across the surface of a sample. As the name suggests, the magnetic field sensor is based on a Superconducting QUantum Interference Device (SQUID). It essentially consists of two superconductors forming a loop, separated by two thin insulating layers, creating a pair of Josephson junctions (JJ). An external flux Φ_{ext} threading through the loop will induce an opposing, circulating screening current I_s in the superconducting ring and due to the quantisation of the magnetic flux, the voltage response of the SQUID will vary periodically as a function of flux. Therefore, in principle, a basic SQUID can directly be used as a periodically varying flux-to-voltage converter [79]. In practice however, the voltage-flux characteristic of a SQUID is commonly linearised by operating in a flux locked loop mode [80]. This increases the measurement sensitivity and reduces noise, as one can operate the SQUID at the steepest part of the V-Φ response. To this end, an additional feedback coil (called a modulation coil) couples to the SQUID and compensates the external magnetic flux in the loop based on the induced screening current [81], ensuring that total flux in the superconducting loop remains constant.

Furthermore, a SQUID can be utilised as a micron-scale ac susceptometer. To do so, a local magnetic field is generated close to the sample surface by applying a low frequency ac current to the so-called field coil (c.f. Fig. 4.26a). In response to the magnetic field, the sample locally generates flux, which is detected by the SQUID pick-up loop. In the case of a superconducting sample, the perfect diamagnetic response due to Meissner screening will lead to a strong expulsion of the external field and consequently a substantial negative susceptibility signal can be detected. This forms the basis of observing the spatially varying superconductivity in microstructured $CeIrIn_5$ lamellae.

Lastly, a further complication typically found in scanning SQUID sensors is the counter wound SQUID design, which features a mirror symmetric SQUID loop with a twist in its centre (c.f. Fig. 4.26a). The counter winding ensures that the net flux coupling into the SQUID loop is dominated by the flux of the sample Φ_{sample}. Since only one pick-up loop (primary side) is in close proximity with the sample, while the second pick-up loop is located approximately 1 mm away and, at the same time, any global background flux variation $\Phi_{\text{background}}$ will couple equally but oppositely into both pick-up loops due to the twist.

The primary pick-up loop (red) and field coil (blue) of the SQUID sensor used in this investigation are shown in Fig. 4.26b. The respective inner/outer diameters are 1.5 μm/1.6 μm for the pick-up loop and 6 μm/12 μm for the field coil, respectively. The superconducting niobium layer (yellow) shields most of the SQUID loop, such that flux from the sample can only permeate the area of the primary pick-up loop, enabling the detection of the local susceptibility response of the sample. To construct a spatially resolved image of the superconducting regions of a lamella, the SQUID sensor is scanned parallel to the sample surface at a height of approximately 1–3 μm (c.f. Fig. 4.26c), while the local susceptibility signal is recorded with a standard lock-in amplifier. By repeating this process at several fixed temperatures, a map of the superconducting temperature distribution can be created.

This concludes the basic review of scanning SQUID microscopy. An extensive review can for instance be found at [82]. I will now turn to the discussion of the spatially modulated superconductivity in microstructured $CeIrIn_5$ devices by starting with the model case of the unstructured lamella.

The local susceptibility of an unstructured $CeIrIn_5$ lamella bonded on a sapphire substrate is displayed in Fig. 4.27a. A large negative susceptibility demonstrates strong Meissner shielding and indicates the regions of superconductivity in yellow. Below 200 mK the majority of the lamella is superconducting. As the temperature is increased, the suppressed T_c along the strained 'a edges' is exceeded prior to the enhanced T_c along the strained 'c edges' of the lamella. As a result, a distinct bow tie

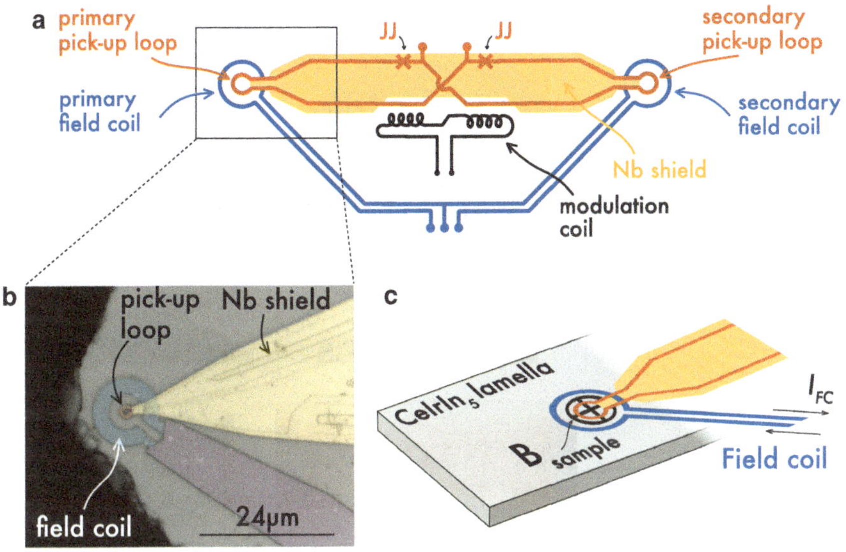

Fig. 4.26 **a** A schematic diagram of a scanning SQUID sensor. The pick-up loop and field coil depicted on the left are mirror symmetric to their counter wound copies on the opposite side of the sensor. The red trace indicates the superconducting loop, containing two Josephson junctions (JJ) marked with 'x'. The majority of the SQUID loop, however, is shielded by a superconducting niobium film (yellow), such that only flux threading the pick-up loops or stemming from the modulation coil (black) couples into the SQUID. The blue trace represents the field coil used for local susceptibility measurements. **b** An image of the primary side of the SQUID sensor. **c** Schematic view of the SQUID scanning operation across the surface of a $CeIrIn_5$ lamella

shaped superconducting pattern develops. As the temperature is further increased, only the areas with the largest strain along the *c*-direction and hence highest T_c support Meissner shielding. Hence, because a side-to-side connecting superconducting path along the c axis is established at higher temperatures than along the a direction ('T_c^{c}' > 'T_c^{a}'), this explains the split superconducting transition observed in our electrical transport measurements. The numerically simulated superconducting regions (in white) at the corresponding temperatures, shown in panel Fig. 4.27b, are in remarkable agreement with the susceptibility measurements, considering that the calculations are performed completely ab initio and do not involve any free tuning parameters. Clearly, the characteristic bow tie shape is recognisable as well. The calculated T_c map presented in panel Fig. 4.27c further validates our understanding of the relation between the superconducting behaviour observed in transport and the microscopic strain situation.

Next, let us discuss the lamella structured in a Montgomery geometry, displayed in Fig. 4.28. A similar bow tie shaped modulation of the superconducting temperature can be observed as the temperature of the lamella is increased, both in the susceptibility measurement as well as in the simulations. The strain along the edges of the central square of the device is evidently the strongest, as it is strained outwardly

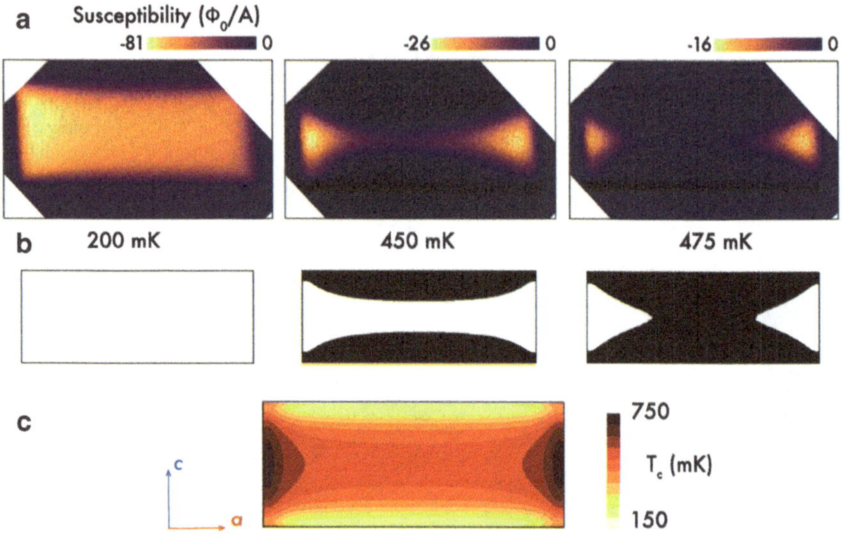

Fig. 4.27 **a** The top row shows the measured susceptibility of an unstructured lamella at 200, 450 and 475 mK respectively. A strong negative susceptibility response (yellow) indicates the regions of superconductivity. **b** The middle row displays the calculated regions of superconductivity (in white) at the same temperatures. **c** The T_c maps of an $a-c$ $CeIrIn_5$ based on COMSOL simulations. The data presented in this figure were measured and analysed by G. Ferguson

at its corners by the outer contacts. The agreement between the measurements and calculations is again astonishing and gives us confidence in the qualitative results of the COMSOL simulations.

Last but not least, the lamella in the modified Montgomery geometry in which the outer contacts connect to the middle rather than to the corners of the central square is displayed in Fig. 4.29. The placement of the contact joints drastically changes the strain situation for the central square, as evidenced by the local susceptibility response (c.f. Fig. 4.29a). Superconductivity initially develops around 550 mK in local regions where the strain along the c axis is the highest. Upon decreasing the temperature, a peculiar spatial variation of T_c develops, characterised by two crescent shaped regions of suppressed superconductivity in the contact pads. This demonstrates how variable the modulation of T_c can be, depending on the design details of the microstructure. A direct comparison of the susceptibility data with the COMSOL simulations (Fig. 4.29b) reveals that the overall temperature scale of the parameter free calculations is distorted relative to the SQUID measurements. In principle, this could be addressed by introducing a phenomenological scaling function which maps $T_c^{\text{calc}} \rightarrow T_c'$. However, in practice we may not just be concerned about the absolute T_c values, but also are interested in the relative evolution of the spatial T_c modulation, which indeed is qualitatively well reproduced as evidenced by the T_c map (Fig. 4.29c). Hence, by appropriately microstructuring a lamella, the potential energy landscape of a sample can be tuned accordingly via the biaxial dif-

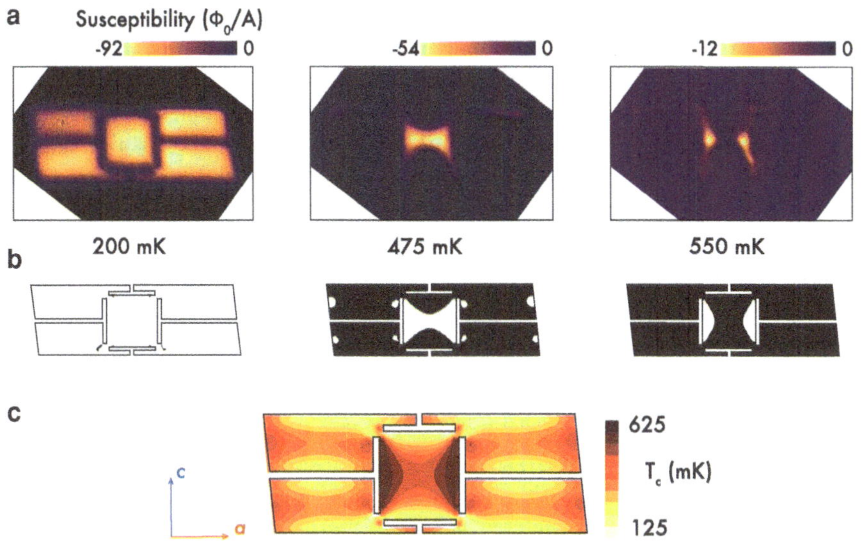

Fig. 4.28 **a** Local susceptibility of a Montgomery device, measured at three temperatures by scanning SQUID microcopy. **b** The results of the COMSOL simulation clearly show how the areas calculated to exhibit superconductivity at the corresponding temperatures (in white) are in remarkable agreement with the measurements. **c** The simulated T_c map of the Montgomery device. The data presented in this figure were measured and analysed by G. Ferguson

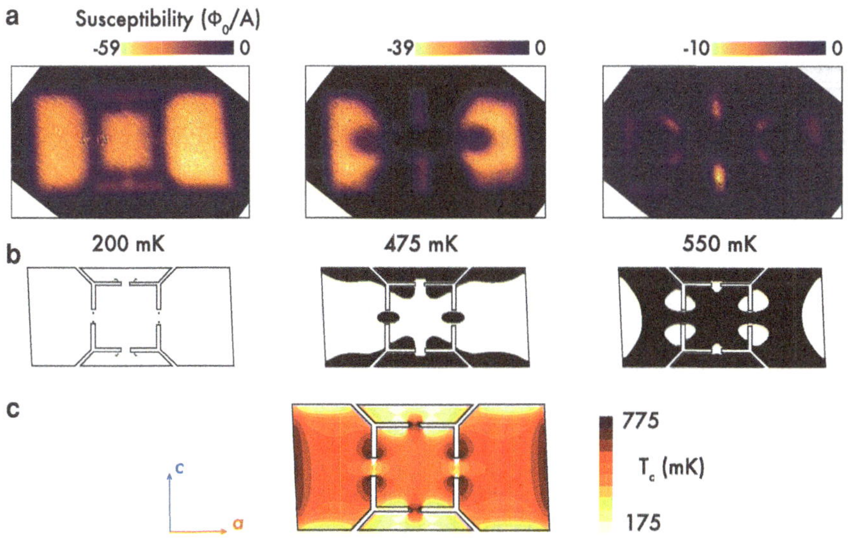

Fig. 4.29 **a** Local susceptibility of a modified Montgomery device, measured at three temperatures by scanning SQUID microcopy. **b** The results of the COMSOL simulation show how the areas calculated to exhibit superconductivity at the corresponding temperatures (in white) are in good qualitative agreement with the measurements. **c** The simulated T_c map of the modified Montgomery device. The data presented in this figure were measured and analysed by G. Ferguson

ferential strain of the substrate. For instance, in the modified Montgomery sample, the enclosed region of suppressed superconductivity acts as a vortex trap, as that is the position of the locally lowest superfluid density.

In conclusion, we have seen that the differential thermal contraction mismatch between an *a*-*c* $CeIrIn_5$ microstructured lamella and a sapphire substrate bonded together leads to an intricate spatial strain modulation within the lamella. Due to the strain sensitivity of T_c, this in turn results in the development of complex spatial variation of the superconducting transition temperature, which can be locally suppressed or enhanced depending on the c/a ratio.

4.5 Conclusion

The results of the scanning SQUID investigations in corroboration with finite element simlulations clearly identify substrate strain as the underlying cause for anisotropic superconductivity in microstructured samples. This, however, raises the question of the origin of the anomalously high transition to a zero-resistance state observed in $CeIrIn_5$ bulk samples. Superconductivity is typically observed below $T_c^{\mathrm{res}} \approx 1.2\,\mathrm{K}$ [21, 45], i.e. about three times higher than the bulk transition observed with thermodynamic probes $T_c^{\mathrm{bulk}} \approx 0.38\,\mathrm{K}$. To unravel this long-standing mystery, we have investigated the superconducting transition in bulk samples of the same high-quality single crystals, which did not show signs of filamentary superconductivity when microstructured.

4.5.1 Macroscopic Samples

We have prepared four transport samples from high-quality single crystals, by wire sawing them into rectangular transport bars along the *a* or *c* direction. Typical dimensions are $300 \times 300 \times 1500\,\mu\mathrm{m}^3$. The samples were contacted by a combination of argon etching and gold sputter deposition (100 nm) onto the contact areas. Silver wires with 25 μm diameter were attached using silver epoxy. To prevent any extrinsic stress, the samples were held in place by the four silver wires alone and not glued onto a substrate.

The zero-resistance transitions are displayed in Fig. 4.30a. Clearly, the T_c values are enhanced, with $T_c^c \approx 0.9\,\mathrm{K}$ and $T_c^a \approx 1.4\,\mathrm{K}$. In a next step, the samples were dipped into 11% hydrochloric acid (HCl) for 10–15 min until a change in surface colour/reflectivity was clearly visible by eye, c.f. panel Fig. 4.30b. This leaves the normal state resistivity unchanged, but the superconducting transition drops close to the thermodynamically expected value. Our observation is consistent with the interpretation that the enhanced T_c in the non-etched samples are a result of surface strain induced by the sample preparation with a wire saw. Upon etching, the strained

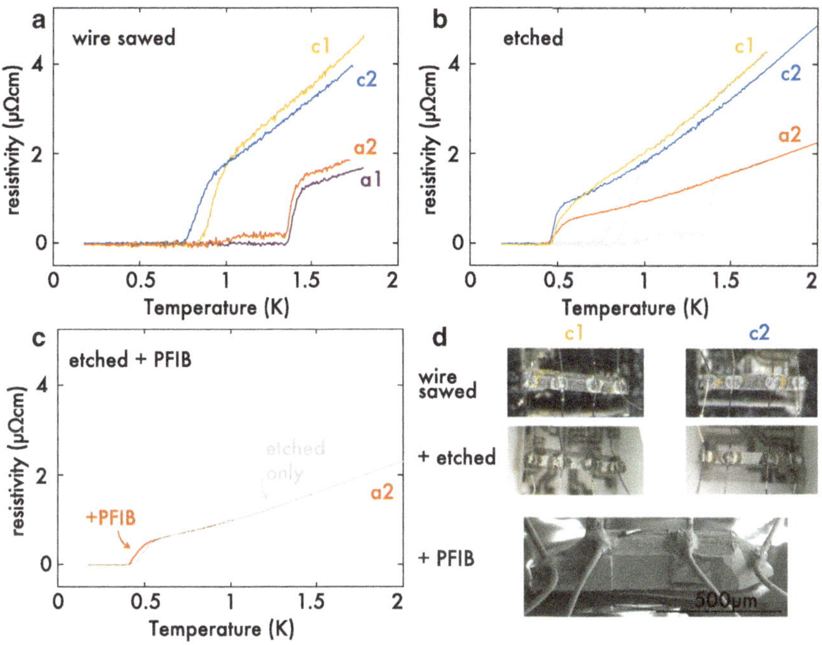

Fig. 4.30 **a** Four macroscopic samples of $CeIrIn_5$ (shown in d) were cut into transport bars by means of a wire saw. All bars showed an elevated T_c. In particular, bars cut along the a direction show the highest T_c observable in $CeIrIn_5$. **b** Upon etching away the damaged surface of the macroscopic samples with HCl, all samples showed a transition into a zero-resistance state close to the thermodynamically expected value. (Sample a_1 was not measured post-etching, because the contacts detached in the process). The grey curves show the data from A in order to compare the normal state resistivity values. **c** The surface of sample a_2 was polished in the xenon plasma FIB after etching. The decrease of T_c to the correct bulk value evidences that the FIB itself does not damage the surface during the milling process. **d** Macroscopic samples, approximately 1.5 mm long, are shown before and after HCl etching. The surface reflectivity has clearly changed. The bottom image displays a SEM image after one side of the sample was polished in the xenon plasma FIB

layer is removed and T_c decreases to its bulk value. This excludes intrinsic strain [31], for instance around defects similar to Ru-inclusions found in Sr_2RuO_4 [83], as the origin of the enhanced transition temperature.

Furthermore, we have taken one of the etched samples and further polished its surface in the xenon plasma FIB, c.f. bottom image in panel Fig. 4.30d. The superconducting transition further decreases towards its bulk value (panel Fig. 4.30c), indicating that even more of the damaged layer is being removed.

These etching experiments clearly demonstrate that surface strain induced by wire sawing or polishing is the reason for the anomalously high resistive transitions. In contrast, FIB machining is a much gentler technique compared to wire-sawing, in terms of inducing strain during sample preparation.

4.5.2 Summary

Our results on microstructured devices as well as macroscopic samples finally resolve the mystery of the anomalous resistive transition in $CeIrIn_5$. We conclude that the superconducting transition in this compound is comparatively sensitive to strain and thus susceptible to differential thermal contraction mismatch or surface induced strain as a result of sample preparation. These experiments demonstrate how involuntary local pressure on the order of 10 kbar can readily be created, even in materials where the response to strain is less dramatic.

In general, the strain accessible using our fabrication route is sufficient to substantially alter the electronic properties of many strongly correlated materials, owing to the small energy scales defining their physics. Having demonstrated the predictive power of our strain simulations based on finite element analysis, we can now begin to envision and simulate future devices exploiting the newly found spatial strain tunability. Strain control of the local correlation length, for instance, is expected to be a viable route to spatially modulate properties beyond superconductivity. In $CeRhIn_5$, for example, controlled strain fields could be used to spatially modulate the magnetic order and enable the conditions for quantum criticality to be introduced locally. Other possibilities include spatial variation in spin and charge density wave systems. On a technological side, strain engineering may offer alternative ways to generate superconducting circuitry in a homogenous metallic device without any physical junctions.

References

1. Landau LD (1957) The theory of a fermi liquid. Sov Phys JETP 3(6):920–925
2. Fisk Z, Canfield PC, Beyerman WP, Thompson JD, Hundley MF, Ott HR, Felder E, Maple MB, Lopez de la Torre MA, Visani P, Seaman CL (1991) Massive electron state in YbBiPt. Phys Rev Lett 67(23):3310–3313
3. Stewart GR (1984) Heavy- fermion systems. Rev Mod Phys 56(4):755–787
4. Fisk Z, Hess DW, Pethick CJ, Pines D, Smith JL, Thompson JD, Willis J (1987) Heavy- electron metals: new highly states of matter correlated states of matter. Science 239:33–42
5. de Haas WJ, van den Berg GJ (1936) The electrical resistance of gold and silver at low temperatures. Physica III(6):440–449
6. Kondo J (1964) Resistance minimum in dilute magnetic alloys. Prog Theor Phys 32(1):37–49
7. Doniach S (1977) The Kondo lattice and weak antiferromagnetism. Phys B 91:231–234
8. Andres K, Graebner JE, Ott HR (1975) 4f- virtual- bound- state formation in $CeAl_3$ at low temperatures. Phys Rev Lett 35(26):1779–1782
9. Bucher E, Maita JP, Hull GW, Fulton RC, Cooper AS (1975) Electronic properties of beryllides of the rare earth and some actinides. Phys Rev B 11(1):440–449
10. Steglich F, Aarts J, Bredl C, Lieke W, Meschede D, Franz W, Schafer H (1979) Superconductivity in the presence of strong Pauli Paramagnetism: $CeCu_2Si_2$. Phys Rev Lett 43(25):1892–1896
11. Tinkham M (1996) Introduction to superconductivity. Dover Publications, Mineola
12. Bardeen J, Cooper LN, Schrieffer JR (1957) Microscopic theory of superconductivity. Phys Rev 106(1):162–164
13. Cooper LN (1956) Bound electron Pairs in a degenerate Fermi gas. Phys Rev 104(4):1189–1190

14. Reynolds, CA, Serin, B, Wright, WH, Nesbitt, LB (1950) Superconductivity of isotopes of mercury. Phys Rev B 78:487
15. Drozdov AP, Eremets MI, Troyan IA, Ksenofontov V, Shylin SI (2015) Conventional superconductivity at 203 Kelvin at high pressures in the sulfur hydride system. Nature 525:73–76
16. Gor'kov LP (1959) Microscopic derivation of the Ginzburg–Landau equations in the theory of superconductivity. Sov Phys JETP 36(6):1364–1367
17. Tsuei CC, Kirtley JR (2000) Pairing symmetry in cuprate superconductors. Rev Mod Phys 72(4):969–1016
18. Izawa K, Yamaguchi H, Matsuda Y, Shishido H, Settai R, Onuki Y (2001) Angular position of nodes in the superconducting gap of quasi- 2D heavy- fermion superconductor $CeCoIn_5$. Phys Rev Lett 87(5):057002
19. Mathur ND, Grosche FM, Julian SR, Walker IR, Freye DM, Haselwimmer RKW, Lonzarich GG (1998) Magnetically mediated superconductivity in heavy fermion compounds. Nature 394:39–43
20. Hegger H, Petrovic C, Moshopoulou EG, Hundley MF, Sarrao JL, Fisk Z, Thompson JD (2000) Pressure- induced superconductivity in quasi- 2D $CeRhIn_5$. Phys Rev Lett 84(21):4986–4989
21. Petrovic C, Movshovich R, Jaime M, Pagliuso PG, Hundley MF, Sarrao JL, Fisk Z, Thompson JD (2001) A new heavy- fermion superconductor $CeIrIn_5$: A relative of the cuprates? Europhys Lett 53(3):354–359
22. Petrovic C, Pagliuso PG, Hundley MF, Movshovich R, Sarrao JL, Thompson JD, Fisk Z, Monthoux P (2001) Heavy- fermion superconductivity in $CeCoIn_5$ at 2.3 K. J Phys: Condens Matter 13:2001
23. Stock C, Rodriguez-Rivera JA, Schmalzl K, Rodriguez EE, Stunault A, Petrovic C (2015) Single to multiquasiparticle excitations in the itinerant helical magnet $CeRhIn_5$. Phys Rev Lett 114:247005
24. Sidorov VA, Nicklas M, Pagliuso PG, Sarrao JL, Bang Y, Balatsky AV, Thompson JD (2002) Superconductivity and quantum criticality in $CeCoIn_5$. Phys Rev Lett 89(15):603–607
25. Knebel G, Méasson MA, Salce B, Aoki D, Braithwaite D, Brison JP, Flouquet J (2004) High-pressure phase diagrams of $CeRhIn_5$ and $CeCoIn_5$ studied by ac calorimetry. J Phys Condens Matter 16:8905–8922
26. Kawasaki S, Yashima M, Mugino Y, Mukuda H (2006) Enhancing the superconducting transition temperature of $CeRh_{1-x}Ir_xIn_5$ due to the strong- coupling effects of antiferromagnetic spin fluctuations: an ^{115} In nuclear quadrupole resonance study. Phys Rev Lett 96:147001
27. White BD, Thompson JD, Maple MB (2015) Unconventional superconductivity in heavy-fermion compounds. Phys C 514:246–278
28. Grin YN, Yarmolyuk YP, Gradyshevsky EI (1979) Crystal structure of the compounds R_2CoGa_8 (R=Sm, Gd, Tb, Dy, Ho, Er, Tm, Lu, Y) and $RCoGa_5$ (R=Gd, Tb, Dy, Er, Tm, Lu, Y). Kristallografiya 24(2):242–246
29. Moshopoulou EG, Fisk Z, Sarrao JL, Thompson JD (2001) Crystal growth and intergrowth structure of the new heavy fermion materials $CeIrIn_5$ and $CeRhIn_5$. J Solid State Chem 158:25–33
30. Haga Y, Inada Y, Harima H, Oikawa K, Murakawa M, Nakawaki H (2001) Quasi- two- dimensional Fermi surfaces of the heavy fermion superconductor $CeIrIn_5$. Phys Rev B 63:060503(R)
31. Bianchi A, Movshovich R, Jaime M, Thompson JD, Pagliuso PG, Sarrao JL (2001) Origin of the zero- resistance anomaly in heavy fermion superconducting $CeIrIn_5$: a clue from magnetic-field and Rh- doping studies. Phys Rev B 64:220504(R)
32. Nair S, Wirth S, Nicklas M, Sarrao JL, Thompson JD, Fisk Z, Steglich F (2008) Precursor state to unconventional superconductivity in $CeIrIn_5$. Phys Rev Lett 100:137003
33. Movshovich R, Jaime M, Thompson JD, Petrovic C, Fisk Z, Pagliuso PG, Sarrao JL (2001) Unconventional superconductivity in $CeIrIn_5$ and $CeCoIn_5$: specific heat and thermal conductivity studies. Phys Rev Lett 86(22):5152–5155
34. Tanatar MA, Paglione J, Nakatsuji S, Hawthorn DG, Boaknin E, Hill RW, Ronning F, Sutherland M, Taillefer L, Petrovic C, Canfield PC, Fisk Z (2005) Unpaired electrons in the heavy- fermion superconductor $CeCoIn_5$. Phys Rev Lett 95:067002

35. Zheng G, Tanabe K, Mito T, Kawasaki S, Kitaoka Y, Aoki D, Haga Y, Onuki Y (2001) Unique spin dynamics and unconventional superconductivity in the layered heavy fermion compound $CeIrIn_5$: NQR evidence. Phys Rev Lett 86(20):4664–4667
36. Allan MP, Massee F, Morr DK, van Dyke J, Rost AW, Mackenzie AP, Petrovic C, Davis JC (2013) Imaging Cooper pairing of heavy fermions in $CeCoIn_5$. Nat Phys 9(8):468–473
37. Zhou BB, Misra S, Eduardo H, Neto S, Aynajian P, Baumbach RE, Thompson JD, Bauer ED, Yazdani A (2013) Visualizing nodal heavy fermion superconductivity in $CeCoIn_5$. Nat Phys 9(8):474–479
38. Lu X, Lee H, Park T, Ronning F, Bauer ED, Thompson JD (2012) Heat- capacity measurements of energy- gap nodes of the heavy- fermion superconductor $CeIrIn_5$ deep inside the pressure-dependent dome structure of its superconducting phase diagram. Phys Rev Lett 108:027001
39. Kittaka S, Aoki Y, Sakakibara T, Sakai A, Nakatsuji S, Tsutsumi Y, Ichioka M, Machida K (2012) Superconducting gap structure of $CeIrIn_5$ from field- angle- resolved measurements of its specific heat. Phys Rev B 85:060505
40. Park T, Bauer ED, Thompson JD (2008) Probing the nodal gap in the pressure- induced heavy fermion superconductor $CeRhIn_5$. Phys Rev Lett 101:177002
41. Oeschler N, Gegenwart P, Lang M, Movshovich R, Sarrao JL, Thompson JD, Steglich F (2003) Uniaxial pressure effects on $CeIrIn_5$ and $CeCoIn_5$ studied by low- temperature thermal expansion. Phys Rev Lett 91(7):076402
42. Dix OM, Swartz AG, Zieve RJ, Cooley J, Sayles TR, Maple MB (2009) Anisotropic dependence of superconductivity on uniaxial pressure in $CeIrIn_5$. Phys Rev Lett 102:197001
43. Nicklas M, Sidorov VA, Borges HA, Pagliuso PG, Sarrao JL, Thompson JD (2004) Two superconducting phases in $CeRh_{1-x}Ir_xIn_5$. Phys Rev B 70:020505(R)
44. Hermannsdörfer T, Heera V, Ignatchik O, Uhlarz M, Mücklich A, Posselt M, Reuther H, Schmidt B, Heinig K-H, Skorupa W, Voelskow M, Wündisch C, Strotzki R, Helm M, Wosnitza J (2009) Superconducting state in a gallium- doped germanium layer at low temperatures. Phys Rev Lett 102:217003
45. Shakeripour H, Tanatar MA, Petrovic C, Taillefer L (2016) Heat transport study of field- tuned quantum criticality in $CeIrIn_5$. Phys. Rev. B 93:075116
46. Miccoli I, Edler F, Pfnür H, Tegenkamp C (2015) The 100th anniversary of the four- point probe technique: the role of probe geometries in isotropic and anisotropic. J Phys: Condens Matter 27:223201
47. Montgomery HC (1971) Method for measuring electrical resistivity of anisotropic materials. J Appl Phys 42(7):2971
48. van der Pauw LJ (1958) A method of measuring specific resistivity and Hall effect of discs of arbitrary shape. Philips Res Rep 13(1):174–182
49. Santos CAM, de Campos A, Luz MS, White BD, Neumeier JJ (2011) Montgomery method Procedure for measuring electrical resistivity of anisotropic materials: a revision of the montgomery method. J Appl Phys 110:083703
50. COMSOL. COMSOL Multiphysics v. 5.3
51. Wüchner S, Keller N, Tholence JL, Flouquet J (1993) Magnetic properties of the heavy- fermion superconductor UPt_3 and URu_2Si_2. Solid State Commun 85(4):355–360
52. Pollini A, Mota AC, Visani P, Pittini R, Juri G, Teruzzi T (1993) Flux dynamics and low-field magnetic properties of the heavy- fermion superconductor $CeCu_2Si_2$. J Low Temp Phys 90:15–53
53. Shishido H, Settai R, Aoki D, Ikeda S, Nakawaki H, Nakamura N, Iizuka T, Inada Y, Sugiyama K, Takeuchi T, Kindo K, Kobayashi TC, Haga Y, Harima H, Aoki Y, Namiki T, Sato H, Onuki Y (2002) Fermi surface, magnetic and superconducting properties of $LaRhIn_5$ and $CeTIn_5$ (T: Co, Rh and Ir). J Phys Soc Jpn 71(1):162–173
54. Clem JR (1989) Phenomenological theory of magnetic structure in the high- temperature superconductors. Phys C 162–164:1137–1142
55. Settai R, Shishido H, Ikeda S, Murakawa Y, Nakashima M, Aoki D, Haga Y, Harima H, Onuki Y (2001) Quasi- two- dimensional Fermi surfaces and the de Haas- van Alphen oscillation in both the normal and superconducting mixed states of $CeCoIn_5$. 13:627–634

56. Schoenberg D (2009) Magnetic oscillations in metals, 1 edn. Cambridge University Press, Cambridge
57. Onsager L (1952) Interpretation of the de Haas- van Alphen effect. London Edinburgh Dublin Philos Mag J Sci 43(344):1006–1008
58. Welser J, Hoyt JL, Gibbons JF (1992) NMOS and PMOS transistors fabricated in strained silicon/relaxed silicon- germanium structures. IDEM Tech Dig 1000–1002
59. Takeuchi T, Inoue T, Sugiyama K, Aoki D (2001) Magnetic and thermal properties of $CeIrIn_5$ and $CeRhIn_5$. J Phys Soc Jpn 70(3):877–883
60. Simon NJ (1994) Cryogenic properties of inorganic insulation materials for ITER magnets: a review. Technical report
61. Ekin JW (2006) Experimental techniques for low- temperature measurements. Oxford University Press, Oxford
62. Shishido H, Settai R, Harima H, Onuki Y (2005) A drastic change of the fermi surface at a critical pressure in $CeRhIn_5$: dHvA study under pressure. J Phys Soc Jpn 74(4):1103–1106
63. Knebel G, Aoki D, Braithwaite D, Salce B, Flouquet J (2006) Coexistence of antiferromagnetism and superconductivity in $CeRhIn_5$ under high pressure and magnetic field. Phys Rev B 74:020501(R)
64. Paglione J, Ho P, Maple MB, Tanatar MA, Taillefer L, Lee Y, Petrovic C (2008) Ambient- pressure bulk superconductivity deep in the magnetic state of $CeRhIn_5$. Phys Rev B 77:100505(R)
65. Johnson SD, Zieve RJ, Cooley JC (2011) Nonlinear effect of uniaxial pressure on superconductivity in $CeCoIn_5$. Phys Rev B 83:144510
66. Borth R, Lengyel E, Pagliuso PG, Sarrao JL, Sparn G, Steglich F, Thompson JD (2002) Heat capacity of the heavy fermion superconductor $CeIrIn_5$ under hydrostatic pressure. Phys B 312–313:136–137
67. Nicklas M, Borth R, Lengyel E, Pagliuso PG, Sarrao JL, Sidorov VA, Sparn G, Steglich F, Thompson JD (2001) Response of the heavy- fermion superconductor $CeCoIn_5$ to pressure: roles of dimensionality and proximity to a quantum- critical point. J Phys: Condens Matter 13:L905–L912
68. Fitzgerald EA, Samavedam SB, Xie YH, Giovane LM (1997) Influence of strain on semiconductor thin film epitaxy Influence of strain on semiconductor thin film epitaxy. J Vacuum Sci Technol 15:1048
69. Lyu J, Fina I, Solanas R, Fontcuberta J, Sanchez F (2018) Tailoring lattice strain and ferroelectric polarization of epitaxial $BaTiO_3$ thin films on Si (001). Sci. Rep. 8:495
70. Liu Y, Wang W, Zhang F, Wang Q (2018) Superconductivity in Sr_2RuO_4 thin films under biaxial strain. Phys Rev B 97:224522
71. Androulidakis C, Koukaras EN, Parthenios J, Kalosakas G, Papagelis K, Galiotis C (2015) Graphene flakes under controlled biaxial deformation. Sci Rep 5:18219
72. Frisenda R, Drüppel M, Schmidt R, Michaelis de Vasconcellos S, Perez de Lara D, Bratschitsch R, Rohlfing M, Castellanos-Gomez A (2017) Biaxial strain tuning of the optical properties of single-layer transition metal dichalcogenides. 2D Mater Appl 10
73. Böhmer AE, Sapkota A, Kreyssig A, Bud SL, Drachuck G, Saunders SM, Goldman AI, Canfield PC (2017) Effect of biaxial strain on the phase transitions of $Ca(Fe_{1-x}Co_x)_2As_2$. Phys Rev Lett 118:107002
74. Visscher WM, Migliori A, Bell TM, Reinert RA (1991) On the normal modes of free vibration of inhomogeneous and anisotropic elastic objects. J Acoust Soc Am 90(4):2154
75. Ramshaw BJ, Shekhter A, McDonald RD, Betts JB, Mitchell JN, Tobash PH, Mielke CH, Bauer ED, Migliori A (2015) Avoided valence transition in a plutonium superconductor. Proc Natl Acad Sci 112(11):3285–3289
76. den Toonder JMJ, van Dommelen JAW, Baaijens FPT (1999) The relation between single crystal elasticity and the effective elastic behaviour of polycrystalline materials: theory, measurement and computation. Model Simul Mater Sci Eng 7:909–928
77. Cook RD, Young WC (1985) Advanced Mechanics of Materials, 1st edn. Macmillan Pub Co., New York
78. Dahlerup-Peterson K, Perrot A (1979) Properties of organic composite materials at cyrogenic temperatures. Technical report
79. Clarke J, Weinstock H (1995) SQUID sensors: fundamentals, fabrication and applications. Springer, Berlin

80. Forgacs RL, Warnick A (1967) Digital- analog magnetometer utilizing superconducting sensor. Rev Sci Instrum 38:214
81. Prêle D, Piat M, Sipile L, Voisin F (2016) Operating point and flux jumps of a SQUID in flux-locked loop. IEEE Trans Appl Superconduct 26(2):1600105
82. Kirtley JR, Wikswo JP Jr (1999) Scanning SQUID microscopy. Ann Rev Mater Sci 29:117–148
83. Kittaka S, Taniguchi H, Yonezawa S, Yaguchi H, Maeno Y (2010) Higher- T_c superconducting phase in Sr_2RuO_4 induced by uniaxial pressure. Phys Rev B 81:180510(R)

Chapter 5
Summary and Outlook

The first chapter of this thesis provided thorough guidelines and instructions on how to fabricate FIB devices, in the hope that microstructuring will gain in popularity over the years to come and set the new standard for electrical transport measurements.

By investigating the ultra-pure metal $PdCoO_2$, I found that even extended FIB irradiation does not deteriorate the crystalline quality of the bulk, such that ballistic transport devices can readily be fabricated. Ballistic measurements are of particular interest in this material, as the underlying hexagonal Fermi surface imposes a simple yet non-trivial complication compared to the commonly studied ballistic electron systems in circular Fermi surface materials. In fact, the flat sections of the hexagonal Fermi surface in $PdCoO_2$ lead to a super-geometrical ballistic effect, which restricts the propagation of the ballistic electrons into three main directions. This has far reaching consequences, as I have demonstrated by measuring the in-plane anisotropy in the ballistic regime as well as by. These experiments form the basis of our microscopic understanding of electron transport in the ultra-clean conductor $PdCoO_2$. In particular in view of the proposed hydrodynamic contributions to the flow of electrons at elevated temperatures, it is necessary to have a solid fundamental description of the ballistic transport, to ground future calculations and experiments.

In the heavy fermion superconductor $CeIrIn_5$, microstructured samples have been found to exhibit spatially modulated superconductivity as a result of the inhomogeneous strain profile induced by differential thermal contraction with respect to the substrate. While puzzling at first, the anisotropic superconducting behaviour of $CeIrIn_5$ led to a fruitful collaboration with K. Nowack's group for scanning SQUID microscopy at Cornell University. By combining resistivity and SQUID microscopy measurements with finite element simulations, we demonstrate that controlled microscopic strain offers a new route towards engineering potential landscapes in strongly correlated matter.

M. D. Bachmann, *Manipulating Anisotropic Transport and Superconductivity by Focused Ion Beam Microstructuring*, Springer Theses,
https://doi.org/10.1007/978-3-030-51362-7_5

I hope that over the course of this thesis, I have conveyed a dual message, namely that FIB microstructuring has considerable utility for quantum materials applications, but that, as with any technique, it also has drawbacks. In particular, its applicability to any new material class has to be carefully investigated on a case-by-case basis. The inevitable surface or edge damage layer from FIB processing is typically only tens of nanometers thick, but if it involves a chemical change it can lead to misleading results. In a large number of materials for which this is not a problem, I believe that the technique has almost limitless potential, and I hope that I have transmitted my enthusiasm for seeing that potential explored!

Appendix A
Multichannel AC Transport Measurements

With the ability to fabricate devices with multiple voltage probes along a single current path comes the need for a multichannel transport measurement setup. To this end we either use a cluster of *Zurich Instruments MFLI digital lock-in amplifiers* or a *Signal Recovery multichannel lock-in amplifier model 9210* with 5 dual channel signal processing modules. For a setup using Zurich lock-in amplifiers (LIA), we use home built voltage-to-current converters [1] as well as pre-amplifiers at room temperature to enhance the signal. The Signal Recovery LIA have the capability of measuring their source current across a built-in $10\,\Omega$ resistor, such that it is sufficient to use a simple pre-resistor as a voltage-to-current converter, while monitoring the current flow. Additionally, due to the low voltage noise of the analog inputs ($1.8\,\text{nV}/\sqrt{\text{Hz}}$) the pre-amplifier stage can be omitted. We these setups a noise level of less than $4\,\text{nV}/\sqrt{\text{Hz}}$ can be readily achieved.

For a standard 4-point resistance measurement of a metallic microstructure, the typical measurement currents used are on the order of $10\,\mu\text{A}$. In this case, a $20\,\mu\text{m}$ long bar with a cross section of $3 \times 3\,\mu\text{m}^2$ and a low temperature resistivity value of $1\,\mu\Omega\text{cm}$, yields a voltage drop of $220\,\text{nV}$, which is well above the noise-level. A common measurement frequency that I use is $177\,\text{Hz}$.

For the ultra-high conductivity metal $PdCoO_2$, which has a low temperature resistivity value of only $10\,\text{n}\Omega\text{cm}$, a current of only $10\,\mu\text{A}$ is not sufficient to exceed the noise-threshold, in particular in the non-local measurement geometry of a transverse electron focusing device. Therefore, for the TEF devices a measurement current between $100\,\mu\text{A}$ and $5\,\text{mA}$ was used. In all cases I verified that the voltage response of the device was in the linear regime.

Reference

1. Barber ME (2017) Uniaxial stress technique and investigations of correlated electron systems. PhD thesis, University of St Andrews

M. D. Bachmann, *Manipulating Anisotropic Transport and Superconductivity by Focused Ion Beam Microstructuring*, Springer Theses,
https://doi.org/10.1007/978-3-030-51362-7

The manufacturer's authorised representative in the EU is Springer
Nature Customer Service Centre GmbH, Europaplatz 3, 69115 Heidelberg,
Germany. If you have any concerns regarding our products, please
contact ProductSafety@springernature.com

Printed and bound by CPI Group (UK) Ltd, Croydon, CR0 4YY
30/09/2026
02225515-0012